Systems of Art

Francis Halsall

Systems of Art

Art, History and Systems Theory

PETER LANG
Oxford · Bern · Berlin · Bruxelles · Frankfurt am Main · New York · Wien

Bibliographic information published by Die Deutsche Bibliothek
Die Deutsche Bibliothek lists this publication in the Deutsche Nationalbibliografie; detailed bibliographic data is available on the Internet at ‹http://dnb.ddb.de›.

British Library and Library of Congress Cataloguing-in-Publication Data:
A catalogue record for this book is available from *The British Library*, Great Britain, and from *The Library of Congress*, USA

Cover design: Adrian Baggett, Peter Lang Ltd

ISBN 978-3-03911-073-5

Hochfeldstrasse 32, Postfach 746, CH-3000 Bern 9, Switzerland
info@peterlang.com, www.peterlang.com, www.peterlang.net

Acknowledgements

The following people read or heard as presentations various parts of this book. I thank them for their support and advice: Richard Hooker, David Hopkins, Robert Linsley, Matthew Rampley, Richard Shiff. Tina Fiske, Alexander Kennedy and Dominic Paterson all provided intellectual inspiration. Thanks also to Aoife Leach for her thorough research work for the book. Particular thanks go to Kitty Zijlmans who has been supportive throughout the writing of the book and I am honoured that she has written the foreword. And most of all I am grateful in ways I cannot express for the keen editorial eye, input, advice and above all support of Julia Jansen without whom this book could not have been written.

Foreword

When Francis Halsall asked me to write a foreword to his book *Systems of Art*, I readily accepted. I am very pleased to see that the complex but fascinating systems theoretical way of thinking is gaining ground in art history. This transition was greatly helped by the translation from the German of the book *Art as a Social System* by the social theorist Niklas Luhmann (1927–1998) in 2000.[1] His ideas about how the art system operates as a dynamic network of communicative utterances then became accessible to the English-speaking world and thus set off explorations of what systems theory can mean for the study of art. The emergence and hence expansion of such a discourse is in fact precisely what Luhmann describes as the *autopoietic*, the self-generating operation of a system. When the system 'picks it up' and an utterance – be it a book, an art work, or a theory – incites reactions (more utterances) the discourse will develop. There is no teleological development, fixed direction, or course-plotting, just contingency. Luhmann's social systems do not consist of people, but of communications; they are emerging cybernetic operations which, when they are successful, will induce more communications, and should they fail to do so will die away. Key words are communication and self-production.[2]

I became familiar with this theory of dynamic systems when I was working on my PhD research in the late 1980s. Luhmann had just published his book *Soziale Systeme* (1984),[3] the first of a long series of books (more than 50) – not to mention the hundreds of articles – on

1 Niklas Luhmann, *Die Kunst der Gesellschaft* (Frankfurt/M.: Suhrkamp, 1995). English edition: *Art as a Social System*, trans. Eva M. Knodt (Stanford: Stanford University Press, 2000).

2 See Francis Halsall, 'Niklas Luhmann (1927–1998),' in Diarmuid Costello and Jonathan Vickery (eds.), *Art. Key Contemporary Thinkers* (Oxford/New York: Berg, 2007), pp. 187–190.

3 Niklas Luhmann, *Soziale Systeme. Grundriss einer allgemeinen Theorie* (Frankfurt/M.: Suhrkamp, 1984); English edition: *Social Systems*, trans. John Bednarz and Dirk Baecker (Stanford: Stanford University Press 1995).

the workings of the various systems of society as he distinguishes them: economic, politics, law, religion, science, art. It took me quite a while to grasp this highly abstract theorizing, but it has remained with me right up to the present. In the past few decades, Luhmann's work has had and still has a wide reception in the German-speaking countries in a multitude of scholarly fields, but for art history it had to wait for the English publication of *Art as a Social System* to find the general acceptance it deserves.

The present book does not focus exclusively on the systems theory of Niklas Luhmann; rather, it situates it in the larger context of systems theoretical approaches to thinking about art. Francis Halsall's starting point for this enterprise is the postulated asymmetry between a discursive system and the object it studies, especially in the case of art after modernism. A multitude of contemporary art practices moves far beyond the creation of art objects, deploying all sorts of tangible and less tangible practices and discourses. The systemic qualities apply to both the art works/projects and the systems of display, exchange, and reception. In the latter case, Danto's concept of *artworld* meets up with a systems theoretical re-composition; in the former case, the problem of what an art work represents is exchanged for the question: What can be observed? Besides this, the phenomenon of systems art is scrutinized, for example, the work of such artists as Hans Haacke, Jack Burnham and more recent examples and furthermore the gallery as a complex system of display and preservation. This specific application of systems-theoretical thinking relies upon such foundations as the work of W. Brian Arthur, former Professor of Economics and Population Studies at Stanford University, who pioneered the modern study of positive feedbacks or increasing returns in the economy.

The scope of systems theory is enormous. One of its founders, Ludwig von Bertalanffy (1901–1972), developed a dynamic theory of stationary open systems, the General System Theory, and he is said to be one of the first to apply the methodology of this system to psychology and the social sciences. In sociology, the American sociologist Talcott Parsons (1902–1979) and his structural functionalism was of paramount importance; it was his general theory of action which set Niklas Luhmann on the track of the double contingency of communication which became fundamental to Luhmann's theory. The term and concept of *autopoiesis* was introduced by the Chilean biologists Humberto Maturana and Francisco Varela in 1973 to describe the self-creation of biological, living systems, and Luhmann applied this notion to the workings of social systems: social systems build themselves through communication. Although Luhmann developed his grand theory of social systems to describe how modern society functions, scholars in the field of the Humanities have tended towards the application of Luhmann's theory to the study and understanding of such other fields as art and literature. Especially in the field of literary studies in the German-speaking world, Luhmann's systems theory has struck a chord.[4] As Henk de Berg points out, there are several directions in the application of Luhmann's theory in the field of literary and art studies, ranging from understanding literary texts/art works as moments within the process of the differentiation of systems in society; as a result of the art system's operations; as communication, to the application of the general systems theory – art as a social system. These applications have now also found a wider acceptance in the English-speaking field of art history.[5]

Francis Halsall sees art history as a complex system of observation, differentiation and representation as distinct 'patterns of systemic behaviour', based upon the assumption that the system of art history is a distributed system of representation (see Chapter 6). In the following, I would like to present some ideas on the anti-ontological approach of art focusing not on the concept of representation, but on the concept of selection. This application of Luhmann's theory – also supported by the ideas of the English historian Quentin Skinner[6] – was developed by a group of scholars in the field of (mainly German) literary

4 See Henk De Berg and Johannes J.K. Schmidt (eds.), *Rezeption und Reflexion. Zur Resonanz der Systemtheorie Niklas Luhmanns ausserhalb der Soziologie* (Frankfurt/M.: Suhrkamp, 2000).

5 See Matthew Rampley, 'Systems Aesthetics: Burnham and Others', in: *Vector [e-zine]*, b # 12 (2005); http://virose.pt/vector/b_12/rampley.html See also Matthew Rampley, 'Art as a Social System: The Sociological Aesthetics of Niklas Luhmann', in: *Telos* (forthcoming 2008).

6 Quentin Skinner, 'Meaning and Understanding in the History of Ideas', in *History and Theory*, 8 (1969) pp. 3–53.

studies and one art historian (me) at the University of Leiden in the late 1980s and early 1990s.[7]

The basic assumption of the Leiden group is that an art work or a text is conceived as a communicative act in a very specific, once-only temporal/spatial context – in fact the word context is misleading for it has a connotation of being static whereas in this model it is seen as a dynamic framework which is only constituted in and by the communicative process itself. Hence, a communicative utterance only has meaning within the communicative situation within which it is voiced, bearing in mind that other options were not chosen. These non-selected possibilities form the context against which the chosen option is profiled and conceptualised. Therefore in his theory Luhmann is not concerned with the people who send (the active person) and who receive (the passive person) so that the message becomes the package of information between transacting people; rather his premise is that every communication constitutes its meaning via the negation of specific, other communicative possibilities. Only those which find a 'follow-up,' which gives rise to more communication, are successful; they are the art system's operational 'motors.' In this framework of a differential communicative context, the utterance takes shape and its meaning lays, or rather, it is constituted in and by the communicative process itself – it is generated by that communication. Instead of the one-way sender → receiver model of communication, a threefold process of selectivity is assumed: firstly, the selectivity from an area of possibilities by the 'sender' (that is the *information*); secondly, the thematising of the information by way of an *utterance*; and thirdly, the selectivity by the 'receiver' in *understanding* that an act of communi-

7 See Jos Hoogeveen and Hans Würzner (eds.), *Ideologie und Literaturwissenschaft* (Amsterdam: Rodopi, 1986); Henk De Berg and Matthias Prangel (eds.), *Kommunikation und Differenz. Systemtheoretische Ansätze in der Literatur- und Kunstwissenschaft* (Opladen: Westdeutscher Verlag, 1993); De Berg/ Prangel (eds.), *Differenzen. Systemtheorie zwischen Dekonstruktion und Konstruktivismus* (Tübingen und Basel: Francke, 1995); De Berg/Prangel (eds.), *Systemtheorie und Hermeneutik* (Tübingen und Basel: Francke, 1997); Kitty Zijlmans, 'Kunstgeschichte als Systemtheorie', in: Marlite Halbertsma and Kitty Zijlmans (eds.), *Gesichtspunkte. Kunstgeschichte Heute* (Berlin: Reimer, 1995) pp: 251–278.

cation has occurred. From this follows that communications are seen as specific locations (position-findings) within (historical) discourses. Within a particular context, a specific position is taken, thereby dismissing or ignoring other possibilities. This, however, happens in the act of communication; the ignored/dismissed context is not a fixed entity but is evoked by and inherent in the communicative utterance. Communications which are successful incite more communications and this contingent process leads to an ever-increasing complexity of communications. This is how systems produce, build, themselves as it were. There is nothing beyond or outside the operations of the system. When art works are regarded as communications, it is not their materiality which is relevant, or its maker, but the work as a communicative act responded to by new communicative utterances, positive or negative, in the form of other art works or as criticism.

Communication is the ruling principle of social systems, and the art system is a specific function of society, just as those of such other social systems as politics, economics, law, religion, science (the latter not entirely covering the broad concept of *Wissenschaft* in German, which refers to the natural and the social sciences as well as the humanities). They all emerged in the process of the functional differentiation of modern society in the course of the eighteenth century. The art system is seen by Luhmann as a *specific* social system because it operates on the basis of a *specific* coded communication which cannot be taken over by any of the other social systems. The code which governs the art system is the binary opposition of beautiful/ugly. It may come as no surprise that this binary code has been the cause of much misunderstanding – in fact a very systemic sequel to this, for many, annoying statement (communications often *do* induce fierce responses). The code beautiful/ugly, however, is not a fixed one, nor is it ontological; it is a set of positions in which the one 'side' (beautiful) means that the utterance is successful for the art system, and the other (ugly) means that it has failed for the art system. It does not mean that the one art work *is* beautiful and the other *is* ugly, but that the one utterance entices more response (be it negative or positive) and the other not. From this point of view, there can never be such a thing at work as a *Zeitgeist*, a period code or any other teleological force driving society. One can never predict which utterance will generate reactions;

the system operates in a highly contingent way. In this respect, the concept of *evolution* is more appropriate than that of *history* – another aspect which is difficult to swallow for art historians.

None the less, the question of what systems theory can bring in art history can be answered in a positive way, and it is also that for which Francis Halsall is aiming in this book. Indeed, he even poses this question in his first chapter. For one thing, a system theoretical look upon art shifts the attention away from a singular art object and its visual appearance to the system and the processes it generates internally. In his concluding chapter, Francis Halsall extends the system theoretical viewpoint to the practice of Art History which is seen accordingly as a complex system of observation. Since the art historical discourse is also communication, it is driven by a similar systemic working in which some utterances and observation will arouse only a little reaction and others a multitude. It is my profound hope that this book will have the latter effect.

Kitty Zijlmans

Contents

Introduction

This book is about systems and about art. In it I explore the implications of using the concept of 'system' within art discourse. This means understanding both art as a system and how art is also a part of other systems. It is my claim that this requires a rethinking of our ideas about both art and art discourse in ways that are both radical and effective.

Recently the artworld has become interested again in the idea of systems, systems theory and systems aesthetics. In the summer of 2005 Tate Modern held an exhibition called *Open Systems: Rethinking Art c.1970*. For the show the curator Donna de Salvo used the concept of 'system' as way of re-historicizing art from a few years either side of 1970. The exhibition was a success[1] for two key reasons. First, it reflected a more general contemporary interest in the art of the period. The high profile retrospectives of Robert Smithson, Dan Flavin and Donald Judd, (amongst others) have demonstrated the recent interest in art of the period. This is also clear in the proliferation of contemporary work that takes its cues from conceptualism, minimalism and institutional critique. Second, 'system' as a theme proved itself to be potentially both a flexible and suggestive strategy of 'rethinking' the period whilst also providing a focus that is effective, rigorous and engaging. It is the implications of this suggestiveness that I unpack in this book.

This book grew out of two central questions that I want to see resolved.

First, I encountered several passing references made to systems theory, cybernetics, the systems thinking approach and so forth but was unable to determine exactly on what terms these vocabularies were being evoked. These references appeared in texts connected to

1 For reviews of the exhibition and catalogue see Francis Halsall, 'Open Systems: Rethinking Art C. 1970', *CAA Reviews* (2005) and 'Open Systems: Rethinking Art C. 1970', *The Art Book*, 13.3 (Aug. 2006), pp. 26–27.

both of my research interests: art after modernism *and* the historiography of writing on art. For example conceptual artists working with systems and seriality, such as Burnham, Haacke, Bochner and Sol-Lewitt, have made reference to 'systems' in their work; and Robert Smithson, who peppered his writings with allusions to 'systems,' included von Bertlanffy's *General System Theory* and Weiner's *Cybernetics* amongst his library of scientific texts.[2] Some critical writing had picked up on this tendency (for example Lawrence Alloway's short paper, 'Network: the Artwork Described as a System,'[3] (1972) and in particular Burnham's writings).[4] But, as the Tate show emphasised, this can be observed as a now passed historical moment, interconnected with the prevalence of conceptual art and the 'Dematerialisation of the Art Object' in the late 1960s and early 1970s.[5] In addition, in 1984 Michael Ann Holly had made a passing reference to von Bertalanffy's *General System Theory* in her comparison of the art

2 See Ann Reynolds, *Robert Smithson: Learning from New Jersey and Elsewhere* (Cambridge, MA: MIT Press, 2003) for the appendix: a catalogue of Smithson's entire library.

3 Lawrence Alloway, 'Network: the Artwork Described as a System', *Artforum* (1972), reprinted in Alloway, *Network: Art and the Complex Present* (Ann Arbor: UMI Research Publishing, 1984).

4 For example Burnham wrote, '[The] cultural obsession with the art object is slowly disappearing and being replaced by what might be called 'systems consciousness.' Actually, this shifts from the direct shaping of matter to a concern for organizing quantities of energy and information.' In Jack Burnham, *Beyond Modern Sculpture: The Effects of Science and Technology on the Sculpture of This Century* (New York: George Braziller, 1968); see also: Jack Burnham: 'Systems Esthetics' *Artforum*, Vol. 7, No. 1 (September 1968) pp. 30–35, http://cadre.sjsu.edu/art103/spring98/readings/system.html (Dec. 2003); 'Real Time Systems,' *Artforum*, Vol. 8, No. 1 (September 1969) pp. 49–55; 'The Aesthetics of Intelligent Systems,' in *On the Future of Art* (New York: Viking Press, 1970), pp. 95–122.

5 Key exhibitions that exemplify the theme of conceptualism as it can be applied to 'systems' and Lucy Lippard's historical conception of the dematerialised art object are: *Information* (July, 1970, MOMA) and *Software: Information Technology – Its New Meaning for Arts* (1970, Jewish Museum, NYC). See Lucy Lippard, *Six Years: the Dematerialization of the Art Object* 1966–1972 (Berkeley: University of California Press, 1997 [1974]).

historians Riegl and Panofsky.[6] She was later to remark to me that she had done so because at a certain time systems theory was very fashionable, being vigorously discussed across a wide variety of disciplines, and that the implications seemed very exciting.[7]

However, if it seems that this appetite for the application of even a very general systems thinking approach within artistic, cultural or historical analyses was something of an historical curiosity rather than a major dominant branch of contemporary thinking then this is only within the English-speaking worlds of art and art discourse. In German speaking academia, for example, systems theory is well-established in its application within sociology; this is centred particularly around the work and reception of Niklas Luhmann. To the contrary in English academia, and art discourse in particular, Luhmann has provoked little interest; and there has still been no systematic attempt in English to apply his systems theoretical approach to thinking about art.[8] This is undoubtedly a reflection of the fact that he was, until very recently, not translated in English (*Social Systems* was translated in 1997 and *Art as a Social System* in 2000) and his death in 1998. Independently of Luhmann, and outside the humanities, however, the discourse of systems thinking itself is a burgeoning discipline with applications as diverse as economics, evolutionary biology, management, and so forth. The area of application that has proved particularly exciting is the ongoing dialogue between cognitive science and philo-

6 Michael Ann Holly, *Panofsky and the Foundations of Art History* (Ithaca: Cornell University Press, 1984), pp. 70, 76, 79.

7 Michael Ann Holly in conversation with the author at the colloquium: *Framing Art History: Reflections on the Discipline* (A European Science Foundation Exploratory Workshop), Edinburgh College of Art, March 13–15, 2003.

8 Notable exceptions include the inclusion of Luhmann in the new edition of Charles Harrison and Paul Wood (eds.), *Art in Theory 1900–2000* (Oxford: Blackwell, 2000); WJT Mitchell's discussion of Luhmann's concept of Medium in *What do Pictures Want?* (Chicago: University of Chicago Press, 2005); and Francis Halsall, 'Niklas Luhmann', in Diarmuid Costello & Jonathan Vickery, *Art: Key Contemporary Thinkers* (Oxford: Berg Publishing, 2007).

sophy when framed by the discursive strategies of systems, complexity, emergence and so forth.[9]

The second main question that this book seeks to address is whether the radical implications of post modern art practice are lost because they are observed through the critical lens of a historical and critical model which is, at heart, modernist. By raising this question I mean to expand upon Craig Owens' claim:

> The poststructuralist critique could not possibly be absorbed by art history without a significant reduction in its polemical force, or without a total transformation of art history itself,[10]

beyond the linguistic application he proposed for it; that is, to include art practice as part of that critique. In other words I work from the assumption that art after modernism challenges us to find new critical paradigms by which to account for it. The central argument of this book is that systems theory is such a paradigm.

There are two further points that I wish to make about the adoption of the systems theoretical approach as an art historian. These points also reflect my joint concerns in late 20th-century artistic practice and historiography.

Firstly, I argue that it is particularly effective in dealing with art after modernism, which is characterised by, amongst other things: non-visual qualities; unstable, or de-materialised physicality and an engagement (often politicised) with the institutional systems of support. By prioritising the systems of support over the individual work of art, such an approach is not tied by an umbilical cord of vision to an analysis based on what individual works of art look like.

Secondly, I have identified a tradition within art historical writing – Podro called it the *Critical Historians of Art* – that is known in the German tradition as *Kunstwissenschaft* (the systematic, or rigorous study of art). I have done so both as a means of clarifying what I mean

9 For a recent survey of these debates see: Jean Petitot et al. (eds.), *Naturalizing Phenomenology: Issues in Contemporary Phenomenology and Cognitive Science* (Stanford: Stanford University Press, 2000).

10 Craig Owens, 'Representation, Appropriation and Power', *Beyond Recognition* (Berkeley: University of California Press, 1992).

when I say art history; but also as a means of identifying a tradition within art history of self-reflexivity and systematic investigation of methods and limits. This is something found in non-German scholarship too, as Podro said:

> This German tradition is not altogether isolated; it is not without relation to the writing of Ruskin, Pater and Fry in England, or to Viollet-le-Duc, Taine and later Focillon and Francastel in France, or to that of Croce, Leonello Venturi and others in Italy.[11]

From a systems theoretical perspective it is an interesting question in its own right to ask why the *Critical Historians of Art* have become the dominant mode of historiography (since the 1980s at least). As a discourse it has become, in systems-theoretical terms, 'locked-in' (via positive feedback) which certainly makes it a valid one to engage with.

It is my view that the systems theoretical approach to art discourse places it within an art historical tradition that includes the *Critical Historians of Art*, and is not in opposition to it. In summary, it is not my intention to either attack or defend a straw-man, or flimsy stereotype of what art history is. I am rather, throughout this book, seeking a body of work, a canon, or discursive system, with which to engage.

In what follows I argue that the systems theoretical approach to art discourse is a continuation of this rich and worthy heritage (of finding historical models to match the art under scrutiny) – not a break from it.

11 Michael Podro, *The Critical Historians of Art* (New Haven: Yale University Press, 1982), p. xxi.

Chapter 1: What is Systems Theory?

Introduction

In the chapters following this one I apply the concept of system and a systems approach to thinking about art. Before I can do so, I need to introduce three central systems theoretical definitions in this chapter.

1. What is a system?
2. What is systems theory?
3. Some key vocabularies of the systems theoretical approach

The term 'system' is familiar to most of us. Even someone who never really heard of systems theory or engaged with specific debates of the systems theoretical approach will have come across at least some of the popular uses of the definitions and vocabularies that I outline below. This is because systems proliferate in all aspects of contemporary society. Systems are all around us. We use them every day in the form of, for example: air conditioning systems; computer operating systems; health-care systems and so forth. The term 'system' is widely applied and might also be used to describe social phenomena as diverse as: housing systems; clothing systems; bedding systems; driving systems and so on. The growth of the phenomena of systems has had noticeable implications for the employment market. For example, *The Bureau of Labor Statistics* (U.S. Department of Labor) gives the following statistical analysis of workers considered to be systems analysts:

> As computer applications expand, systems analysts, computer scientists, and database administrators are projected to be among the fastest growing occupations. Systems analysts, computers scientists, and database administrators are expected to be among the fastest growing occupations [corresponding to an increase of 36 percent or more] through 2010. Systems analysts, computer sci-

> entists, and database administrators held about 887,000 jobs in 2000, including about 71,000 who were self-employed... Employment was distributed among the following detailed occupations:
> Computer system analysts: 431,000
> Network systems and data communications analysts: 119,000
> Database administrators: 106,000
> Computer and information scientists, research: 28,000
> All other computer specialists: 203,000[1]

This shows how widespread 'system analysis' is as a descriptive term. Part of the reason for this lies in the usefulness of 'system' as a type of catch-all term to describe complex items that go together and interact. However, there is a more technical definition to which I will turn in what follows.

1. What is a System?

1.1 Defining a System

Systems can be defined in various ways depending upon how they are being used and who is using them. For example within the discourse of communications theory a system is defined in terms of the flows of information.[2] Systems can be analysed in terms of their structural organisation, as a biologist studying organic systems might.[3] Systems have been identified by those working within disciplines as diverse as: physics, biology, chemistry, engineering, economics, sociology and so

1 Bureau of Labor Statistics, *U.S. Department of Labor*, Occupational Outlook Handbook, 2002–03 Edition, *Systems Analysts, Computer Scientists, and Database Administrators* at: http://www.bls.gov/data/ home.htm (25/11/2007).

2 The classic formulation of information theory was presented by Claude Shannon in 1948 at Bell Labs. The text of his original paper is expanded upon in: Shannon and Weaver, *The Mathematical Theory of Communication* (Chicago: University of Illinois Press, 1963 [1998]).

3 The classic work on a systems theoretical approach to biology was put forward by Gregory Bateson, *Steps to an Ecology of Mind* (New York: Ballantine, 1972).

forth. It has also been used within theories and strategies of management.[4] As will be discussed in further detail below, it is a strength of systems theory that it can accommodate such wide-ranging interests within the vocabularies of a unified discourse. This is because it looks for systems in a diversity of observed phenomena.

Broadly speaking a system is a set of elements integrated with one another to such an extent that they form a recognisable and coherent whole. In addition, this recognisable and coherent whole performs some type of recognisable function. Thus, in general a system is a collection of components that by virtue of its organisation and function, becomes *meaningful*[5] in its own right. As Boulding observed: 'The broadest possible definition of a system is that it is anything that is not chaos. We could turn the pattern around and define a system as any structure that exhibits order and pattern.'[6]

However, a mere aggregation of arbitrary elements on its own does not constitute a system. It needs to be a collection of the right type of parts in the right type of order. For example a drawer full of socks is not a system. It is, instead, merely an aggregate of its parts and will not display the requisite systemic behaviour. In order for a collection of components to be considered a system the following three criteria need to be present:

a. There must be some kind of meaningful order to their arrangement. In other words they need to be organised in a meaningful way.

b. The order of this organisation must enable the performance of an identifiable function.

4 This is exemplified by the work of the Complexity Research Programme at the London School of Economics, an interdisciplinary research and teaching initiative. For further details see the Complexity Research Programme website: http://www.psych.lse.ac.uk:16080/complexity/index.html (11/11/2007). The application of the systems theoretical approach to town planning and social management received a thorough articulation in George Chadwick, *A Systems View of Planning* (Oxford: Pergamon, 1971).

5 A system is identified as meaningful when it has a recognisable structure and when that structure contributes to the functioning of the whole.

6 K.E. Boulding, *The World as a Total System* (Beverly Hills: Sage, 1985), p. 9.

c. The individual components of the system need to be able to perform functions which contribute to the systemic operations of the whole system.

For example a collection of neurons meaningfully ordered in a system of a neural network is able to perform functions that none of the single neurons can perform; (on the contrary there is nothing that a drawer of socks can do – no matter how sophisticated the ordering of the socks – that a single pair of socks cannot do. This is because the individual socks as well as the collection of socks, cannot perform autonomous functions; and the drawer on its own cannot function autonomously either).

For a collection of components to be called a system (either simple, complicated or complex) the elements must be organised in a systemic way. This is to say that not only the system's function be considered but also that the organisation of its elements plays an equally significant role in identifying and defining whether a collection can be called a system. The identification of such systemic organisation requires that the elements are interconnected in such a way that the functions of the individual parts will have an effect on the functioning of the whole system. Likewise the functioning of the system will affect the functions of elements within the system. Within the sock drawer the socks do not function in a manner related to the drawer of socks; instead their 'sock-ness' is autonomous from their location in either the drawer or on the foot of the owner. Likewise the drawer does not function according to the socks contained within it and would function no differently if it contained other items of clothing such as tee-shirts.[7]

7 Von Bertalanffy has defined this in the following terms: 'In dealing with complexes of "elements", three different kinds of distinction may be made – i.e., 1. according to their number; 2. according to their species; 3. according to the relations of elements.' Bertalanffy, *General Systems Theory* (New York: Braziller, 1968). In the example of the sock-drawer: 'number' corresponds to the amount of socks; 'species' refers to the items of clothing (in this case socks); and 'relations' refers to the organisation of the socks in the drawer.

Ludwig von Bertalanffy, who is widely seen as the originator of the systems thinking approach[8] gave the following, more technical, definition:

> A system can be defined as a complex of interacting elements. Interaction means that elements, p, stand in relations, R, so that the behaviour of an element p in R is different from its behaviour in another relation R′. If the behaviours in R and R′ are not different, there is no interaction, and the elements behave independently with respect to the relations R and R′.[9]

Further, von Bertalanffy stated that a system might also be defined mathematically as a series of simultaneous differential equations. In doing so a set of mathematical criteria is provided for modelling the system. The set of conditions of the system's operation and how this will change over time can therefore be qualified as is demonstrated, by von Bertalanffy in the following sequence of differential equations:[10]

$$\frac{Q_1}{dt} = f_1 (Q_1, Q_2, \ldots Q_n)$$

$$\frac{dQ_2}{dt} = f_2 (Q_1, Q_2, \ldots Q_n)$$

$$\cdots\cdots\cdots\cdots\cdots\cdots\cdots\cdots$$

$$\frac{dQ_n}{dt} = f_n (Q_1, Q_2, \ldots Q_n)$$

8 Often called the 'founding father' of systems theory von Bertalanffy established the systems-theoretical approach in 1968 with the publication of *General System Theory*. As a developmental biologist he argued that this approach could use the biological metaphors of adaptation and evolution to explain the behaviour of a wide variety of isomorphic 'systems'.

9 Ibid., p. 55.

10 'For illustration, we choose a system of simultaneous differential equations. Denoting some measure of elements, Pi (i = 1, 2, …n), by Qi , these, for a finite number of elements and in the simplest form will be of the form [of the equation given above]. Change to any measure Qi therefore is a function of all Qs from Qi to Qn; conversely, change of any Qi entails change of all other measures and of the system as a whole' (Ibid., p. 56).

Another definition of systems was provided by Angyal. Like von Bertalanffy he is another figure of historical significance for the birth of systems thinking as a separate discipline.[11] From within a sociological perspective Angyal applied the concept of systems to the behaviour of humans in their environments. He claimed that the:

> Structure of wholes is perhaps amenable to logical treatment after all, that, though it may not be described in terms of relations, it may be described in terms of some more adequate logical unit, representing an entirely different logical genus. Here the attempt will be made to demonstrate that there is a logical genus suitable to the treatment of wholes. We propose to call it a system.[12]

Angyal's conception of 'system' as an entity of structural organisation which is conceptually distinct from its component parts raises another key aspect of systems in general, of central importance to my arguments in the book, that that they are understood not only as combination of component parts but also as a combination of component parts *that can only be understood in terms of its entirety*. In other words the system cannot be accounted for by considering merely the sum of the individual elements. Instead, because the organisation and function in the system as a whole is crucial for their operation, a holistic approach is required.

1.2 Identifying a System

A system, in order to be identified as such, must have coherence and must be recognisable in its own right. This sense of the individual identity of a system is related to its function in so far as the system's

11 Angyal was another key figure in providing clear definitions of 'system' for the nascent discipline. His scientific definitions of 'system' are outlined in: A. Angyal, *Logic of Systems* (London: Penguin, 1969). As a psychologist and psychotherapist Angyal discussed the epistemological implications of systems-thinking with particular reference to the human subject as an extension of his earlier work which focused on the relationship between the psychic system (Luhmann) and its environment. See A. Angyal, *Foundations for a Science of Personality* (Boston: Harvard University Press, 1941).

12 Ibid., pp. 243 ff.

identity is constituted by its performing a recognisable function (or variety of functions) over time. It is by virtue of its function (or functions) that the system acquires and maintains both a clear identity and distinguishable boundaries. It can thus be differentiated both from its environment and other systems. Further, through the use of function(s) (rather than structure) as the criteria for identity, the system can retain its recognisable and distinct identity over time even though its structure may have adapted and evolved. For example a synapse within a neural network cannot be identified as a system because it cannot be functionally differentiated from its environment of the network within which it functions. The neural network itself, on the other hand, can be identified as a system because it has its own coherent and differentiated identity in so far as it will be seen to perform the same recognisable function over time even as its structure evolves. It can also be differentiated from other neural networks that function in different spatio-temporal conditions.

Further a system (or sub-system) might be identifiable in its own right but still be functionally subordinate to a larger system. This would happen in the case of sub-systems which operate within larger systems. Such subordination need not compromise the identity of the sub-system which could still be identified as a system in its own right. This would be possible providing that the system meets the criteria given above of demonstrating a recognisable function over time. An example of this would be the human subject, itself a complex system, yet one which circulates within and is hence subordinate to larger social systems.

1.3 Linear and non-linear systems

Providing a set of qualifying mathematical criteria, as von Bertalanffy does, is necessary if the system is to be modelled conceptually rather than physically. This is of particular significance when the system is modelled by computer as a virtual system (and given the complexity of such systems this is often a necessity).

By using such a conceptual model the nature of the system can be articulated in terms of the types of differential equations which are required for its modelling. For example a simple system such as an environment whose temperature is regulated using a thermostat (which uses negative feedback to regulate temperature of a system within set parameters) can be mathematically modelled using several linear differential equations which give the graph shown below.

The differential equations will be linear because such a system will behave in a predictable way. This means that given an input a predictable output will be generated and this output will be of the same value every time that input is used. Further, given the predictability of the system a specific output can be predicted for *any* possible input introduced into this simple, linear system.

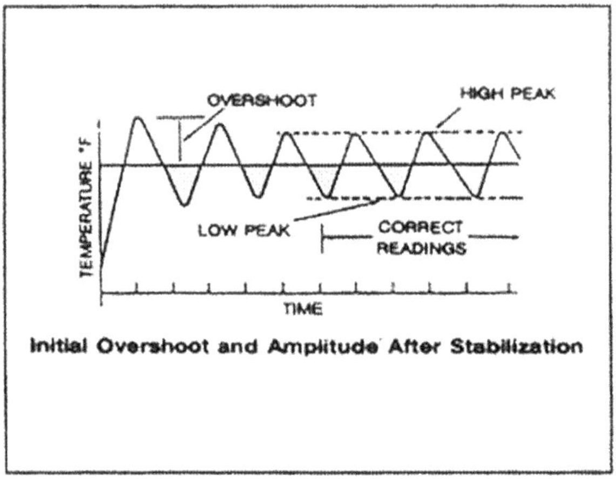

Image of the operations of a thermostat from the American Appliance website[13]

A complex, or non-linear system, on the other hand, behaves differently, and it is these complex or non-linear systems that form the basis of the examples of complex systems used throughout this book. Systems of this type are unlike linear systems because there are unpredictable aspects to their behaviour and they do not display the quality of simple predictability that is found in linear systems. Complexity refers to the fact that systems will not always behave in the same way

13 http://www.americanappliance.com/service_pages/general_tips/range_adjustment.htm (20/03/2007).

given the same input. If such a system is to be modelled (and often the complexity of the system means that it cannot be modelled in its entirety) then the differential equations used in the modelling will also be non-linear. Coveney and Highfield have given some examples of what complex and non-linear systems are like:

> In general, non-linearity produces complex and frequently unexpected results. Irreversibility and nonlinearity characterise phenomena in every field of science: the complexity of the markings on a butterfly's wings, a leopard's spots, the shape of a spleenwort fern, the whorls of a little green alga called the Mermaid's cap, and the rhythms in living systems such as the palpitations of a heart, the firing of nerve cells within the brain, and so on. Related but more subtle, chaotic forms of complexity also arise from non-linearity, including apparently random weather patterns, the outbreak of flu epidemics, and the spread of information and ideas.[14]

2. What is Systems Theory?

Generally speaking, systems theory is the study of systems. It is an umbrella term that includes a wide variety of studies. However the arguments in this book are concerned primarily with study and theory of *complex* systems. The closest historical precedent for the study of complex, non-linear systems as it is understood today is found in the study of cybernetics. The term 'cybernetics' was coined in 1948 when Norbert Weiner, a professor of mathematics at MIT, published his book called *Cybernetics*. Weiner defined his term in the following manner:

> We have decided to call the entire field of control and communication theory, whether in the machine or in the animal, by the name Cybernetics, which we form from the Greek [kybernos] or steersman.[15]

14 Peter Coveney and Roger Highfield, *Frontiers of Complexity* (London: Faber and Faber, 1995), pp. 9–10.

15 Norbert Weiner, *Cybernetics or control and communication in the animal or machine* (Cambridge, MA: MIT Press, 1948).

With his book *Cybernetics* Weiner became a pioneer of the systems theoretical approach. However he did not consider himself an originator of a new way of thinking but instead traced his thought back to the Industrial Revolution. The emergence of the industrial world in the 19th century brought with it, he argued, systemic modes of thought. He identified the analysis of systems emerging from the creation of systems of industrial production and the subsequent need for ways of accounting for them. According to Weiner the first paper of note to be published on feedback mechanisms within industrial systems was published by Clerk Maxwell in 1868. Further, Weiner argued, the modern world as the product of the industrial revolution was comprised not only of industrial systems but also of other social systems that emerged at the same. Such social systems included class systems marked by the emergence of the bourgeoisie as the dominant political force in the western world; national systems marked by the emergence of the nation state; political systems created to reflect these new class and national systems; transport systems marked by the emergence of railway networks; communication systems evidenced by the appearance of newspapers and the telegraph and other forms of mass media; and so forth.

Weiner's project in the establishment of cybernetics was to identify a vocabulary through which a general theory of scientific, social *and systemic* phenomena could be articulated. He therefore sought a language by which to describe the:

> Essential unity of the set of problems centring about communication, control and statistical mechanics, whether in the machine or in living tissue.[16]

2.1 Systems-Thinking and Interdisciplinarity

The 'essential unity' to which Weiner referred demonstrates his intention to identify a shared vocabulary by which to talk about different systems and thus provide common conceptual and mathematical models to be applied across disciplines. This call for interdisciplinarity repre-

16 Ibid., p. 19.

sents a key feature of a systems-theoretical approach. It was later taken by Ludwig von Bertalanffy, the father of modern systems theory, as the defining feature of what he was to name in 1968 a 'general system theory'. As is discussed in further detail below, with this theory von Bertalanffy attempted a project similar to that of Weiner; namely the identification of what he called 'isomorphisms', or structural similarities, shared by different systems.

Like the study of cybernetics the general system theory of von Bertalanffy is interested in different organisations of systems as well as questions about their functions and outputs. Equally it is also concerned with identifying systems; that is in distinguishing systems both from their environment and from other structures that are not systems. This process of differentiation entails the observation of phenomena in terms of systemic input, output and behaviour, and then the subsequent evaluation as to whether systems are present. Areas of research where the systems thinking approach has been meaningfully applied and systemic behaviour identified include meteorology, economics, neuroscience, biology, chemistry, physics, sociology and anthropology. It is my intention in the following chapters to add art history to this ever growing list of applications.

2.2 How Does General System Theory work? – The Operations of Systems Theory

Regardless of its application in the various areas of study there is a unifying feature of the systems theoretical approach. This is that clearly there can be no systems theory without a concept of system. In 1968, in *General System Theory*, Ludwig von Bertalanffy recognised the need for such a plural concept in the systems theoretical approach and stated that it should operate according to the central concept of 'system'. He wrote:

> Thus, there exist models, principles, and laws that apply to generalized systems or their subclasses, Irrespective of their particular kind, the nature of their component elements, and the relations or 'forces' between them. It seems legitimate to ask for a theory, not of systems of a more or less special kind, but of a uni-

> versal principles applying to systems in general... In this way we postulate a new discipline called General System Theory. Its subject matter is the formulation and derivation of those principles which are valid for systems in general... A consequence of the existence of general system properties is the appearance of structural similarities or isomorphism in different fields. There are correspondences in the principles that govern the behaviour of entities that are, intrinsically different.[17]

Broadly speaking, general system theory operates through the analyses of 3 types of systems. These are:

a. Natural Systems. Those systems said to occur naturally. These include such systems as an ant colony, a weather system or a cognitive system such as the brain.

b. Artificial Systems. These are non-natural systems such as economic markets, or models of artificial intelligence which are, in the broadest sense the product of human activity.

c. Epistemological systems. These are systems of knowledge by which the world is represented. Whilst these are closely related to artificial systems (in so far as epistemological systems are not naturally occurring) they are included here as a separate order of systems because they are discursive systems of conceptual self-reflection. Systems theory itself is such an epistemological system. And, as I argue in the following chapters, so too is art history.

Von Bertalanffy developed general system theory as a way to order the many apparently differing yet what he identified as 'isomorphic' (structurally similar) practices that he had observed in diverse scientific fields. In reflecting back on his own historical legacy just four years later (in 1972) he again asserted the centrality of the principle of 'system' in a general system theory as a way of organising what he had observed. Von Bertalanffy wrote:

> This trend [of organization in complex systems] arising in biology and other disciplines led me to conceive the idea of a general Systems Theory, that is, an interdisciplinary doctrine elaborating principles and models that apply to sys-

17 Bertalanffy, *General System Theory*, p. 32.

> tems in general, irrespective of their particular kind, elements and forces involved.[18]

Evidence of dynamical systems theory 'locking in' around the concept of system proliferates the discourse. As Ackoff observed in 1960:

> The term 'system' is used to cover a wide range of phenomena. We speak, for example, of philosophical systems, number systems, communication systems, control systems, educational systems and weapon systems. Some of these are conceptual constructs and others are physical entities. Initially we can define a system broadly and crudely as any entity, conceptual or physical, which consists of interdependent parts.[19]

Lazlo, another pioneer of the systems theoretical approach wrote in 1972:

> I only suggest that such general systems theories grasp some forms of order in the world which elude other types of theories. Thus general systems theories are one species of 'world hypothesis'; they find their use and justification in the elucidation of otherwise chaotic patterns of events in the natural world. That currently general systems theories represent the most intensely worked field of general theory is not accidental: the systems concepts used in this theory permit a more adequate interpretation of a greater range of phenomena in a more consistent and unitary manner than any other.[20]

More recently Kevin Kelly argued in *Out of Control* (1994), his survey of complex systems and what he subtitles 'the new biology of machines, social systems and the economic world', for a wider application of the systems theoretical approach to a variety of analyses:

> The overlap of the mechanical and the lifelike increases year by year... I call these examples [of complex and auto-catalytic systems from ant colonies, to economies to robot intelligence], both made and born, 'vivisystems' for the lifelikeness each kind of system holds.... The vivisystems I examine in this book are nearly bottomless complications, vast in range, and gigantic in nuance.

18 Ludwig von Bertalanffy, 'Foreword', E. Lazlo, *Introduction to Systems Philosophy* (New York: Gordon and Breach, 1972), pp. xvii.

19 Russell Ackoff, 'Systems, organizations and interdisciplinary research', *General Systems Yearbook*, vol. 5, 1960 (Society for General Systems Research), p. 3.

20 Lazlo, *Introduction to Systems Philosophy*, p. 15.

> From these particular big systems I have appropriated unifying principles for all large vivisystems; I call them the laws of god, and they are the fundamentals shared by all self-sustaining, self-improving systems.[21]

What follows below is a general introduction to some more specific vocabularies of systems theory which will be subsequently applied to a discussion on art in the rest of this book.

3. Some Key Vocabularies of the Systems Theoretical Approach

3.1 Dynamical Systems Theory

Systems come in all shapes and sizes; in different types and multiple forms. And likewise studies of systems also come in different forms. For example the broad term 'systems thinking' includes cybernetics, general system theory, chaos theory, complexity theory and so forth. Given the diversity of concerns, disciplines and practitioners involved there is often a lack of clarity in the application and use of specific vocabularies. This lack of clarity is demonstrated by the example of chaos theory which whilst very fashionable during the 1980s consequently fell out of favour in the 1990s and was replaced by the discourse of complexity theory. The terms 'chaos' and 'complexity' often carry an ambiguity with regards to what theories exactly they are meant to invoke. As Chris Langton (a central figure in chaos theory) observed: 'Chaos and Complexity are chasing each other around in a circle trying to find out if they are the same or different.'[22] These ambiguities are exacerbated because systems thinking is still a relatively young discipline that constantly makes new advancements. For this reason it has not yet had time to clarify its terms and methods or effect a consensus amongst practitioners. So, for example, in the discussions

21 See Kevin Kelly, *Out of Control* (New York: Perseus Books, 1994).

22 Chris Langton in conversation quoted in Roger Lewin, *Complexity: Life on the Edge of Chaos* (London: Phoenix, 1993), p. 10.

on emergent properties reference is made to 'emergentism',[23] 'emergence', 'emergent properties' and 'synergetics'[24] in reference to the same observations about the emergence of phenomena from collective patterns of systemic behaviour.

With regards to my own discussion here I am primarily concerned with those particular types of systems known as *dynamic* systems.[25] Dynamic systems are those systems which demonstrate dynamic patterns of behaviour which is to say that they are systems that change over time. Such dynamic patterns include behaviour such as self-organisation, emergence, complexity and unpredictability (as outlined below). For this reason I will use the term 'dynamical systems-theory', that is, the study of dynamical systems, to refer collectively to the various theories which I draw from in my arguments concerning the application of the systems theoretical approach to thinking about art.

3.2 Closed and Open Systems

A key characteristic of dynamic systems is that they are open systems. Open systems are dynamic and change over time because they are

23 See Jean Petitot et al., 'Beyond the Gap: An introduction to Naturalizing Phenomenology', in Jean Petitot et al. (eds.), *Naturalizing Phenomenology* (Stanford: Stanford University Press, 1999), pp. 1–82.

24 H. Haken, *Synergetics: An Introduction: Nonequilibrium Phase Transitions and Self-Organization in Physics, Chemistry, and Biology* (New York: Springer, 1978).

25 The term 'dynamical systems' comes from the group 'The Dynamical Systems Collective' which was established in the 1970s at the University of California, Santa Cruz to study the nascent discipline of chaos theory. The group was formed by Norman Packard and Doyne Farmer who famously tried to discover a system by which to beat the casino tables in Las Vegas. Further anecdotal evidence of this can be found in Thomas Bass, *The Predictors: How a Band of Maverick Physicists Set Out to Beat Wall Street* (London: Penguin, 2001). The key paper by the members of the collective is Crutchfield, Farmer, Packard, Shaw, 'Chaos', *Scientific American*, 255 (December, 1986) pp. 46–57, in which they discuss chaotic dynamics in the operation of both computers and in intervals between drops of water dripping unpredictably from a tap.

open to influence from their immediate environment. Open systems are differentiated from closed systems which are, conversely, operationally separate from their surroundings.[26]

One can imagine a closed system by imagining two glass jars; on top of one another and separated by a card divider. In the top jar is black smoke (which is heavier than air), and in the bottom jar is fresh air. The contents of the two jars are separated from one another by the divider and from the external environment by the walls of the jars; it is thus a closed system. If the separating divider is removed the smoke will diffuse into the air of the lower glass. After a period of time both jars will be full of an even mixture of smoke and air. At this point the system will have reached a point of equilibrium, which is to say that at this point the air and smoke cannot mix any further and thus the closed system ceases to evolve. In contrast an open system will continue to receive information from its immediate environment and in doing so it will adapt its operations and form accordingly.

The difference between closed and open systems is a central focus of attention for two of the most important studies of systems theory: von Bertanlanffy's *General System Theory* and Prigogine and Stengers' study of complexity, *Order out of Chaos*.[27] In *Order out of Chaos* it is argued that when dynamical systems reach a certain level of complexity they display self-organising (autopoietic)[28] behaviour.

26 The significance of open systems was identified as early as 1950 in von Bertalanffy's paper, 'The Theory of Open Systems in Physics and Biology'. In this paper von Bertalanffy identified that organisms could be discussed in terms of being open systems in so far as they interacted with their immediate environments by 'maintaining themselves in exchange of materials with environment, and in continuous building up and breaking down of their components'. See Ludwig von Bertalanffy, 'The Theory of Open Systems in Physics and Biology', *Science*, vol.111 (1950), pp. 23–29.

27 Ilya Prigogine and Isabella Stengers, *Order Out of Chaos: Man's New Dialogue with Nature* (New York: Bantam, 1984).

28 Autopoiesis is 'the process whereby an organization produces itself. An autopoietic organization is an autonomous and self-maintaining unity which contains component-producing processes. The components, through their interaction, generate recursively the same network of processes which produced them. An autopoietic system is operationally closed and structurally state determined with no apparent inputs and outputs. A cell, an organism, and perhaps a

This principle of self-organisation is a key characteristic of complex systems. It expresses how complexity emerges from systems which are open to influence from their environment(s).

Open systems interact with their environment in two ways. They can receive information from outside their boundaries which subsequently has an effect upon the internal organisation of the system. And they can transmit information to their environment too. Neither process of input or output occurs in the smoke filled jars of the above example.

The distinction between closed and open systems is often invoked by the system theoretical approach as a means to demonstrate the limits of classical physics, as exemplified by the Second Law of Thermodynamics,[29] in dealing with the emergent dynamics of complex systems. As von Bertalannfy noted, classic physics looks at closed systems which do not interact with their environment. These include models as small as an atom or as large as a solar system. The benefit of dealing with these entities as closed systems is that their behaviour can be successfully predicted. This is because if the all the conditions of the system and the way in which that system will perform are known it can be modelled as if it were isolated from its environment.

Prigogine provided the following mathematical definition of the Second Law of Thermodynamics as formulated by Rudolf Clausius with Lord Kelvin some time around 1820.[30] Simply speaking it states that within a closed physical system the disorder or entropy of that

corporation are examples of autopoietic systems'. From Francis Heylighten (ed.), *Web Dictionary of Cybernetics and Systems*, on the website of the *Principia Cybernetica Project* (PCP) at http://pespmc1.vub.ac.be/Default.html (29/11/2007).

29 Dynamics is a branch of mechanics which concerns itself with the study (and mathematical modelling) of the behaviour and action of physical bodies when acted upon by physical forces. It also studies the changes in objects' motions produced by such forces.

30 Ilya Prigogine, 'Thermodynamics of Evolution'. *Physics Today* (November, 1972), pp. 23–28 and (December, 1972), pp. 34–44.

system will increase over time until equilibrium is reached and the system is in the state of maximum entropy.[31]

As a further development of this theme, von Bertalanffy, extrapolated by providing a conceptual congruence between the concept of entropy and the concept of information. In these terms, information is defined accordingly, as (in Sayre's terms) 'decrease in uncertainty, or more directly as an increase in probability'.[32]

In his use of the concept of information Von Bertalanffy appeals to the concepts of order and the lack of order (or disorder) as a means of describing the state of a system at any one time. A disordered system is one which demonstrates minimal organisational information; and likewise an ordered system manifests the information of the systems' structural organisation.

Thus the Second Law of Thermodynamics can be restated using the concept of information. In doing so it will be stated thus: that within a closed system the quantity of entropy (or disorder) will increase until a state of equilibrium is reached; and at this stage the system will contain minimal information. In this formulation a closed system will lose organisational information over time and become increasingly entropic.

$$\frac{dS}{dt} \geq 0$$

An order of entropy can be expressed mathematically as a logarithm of probability because, it is argued, the entropic state is the most probable state for a closed system to be in. In the example of a closed system given above it is highly unlikely that the smoke and air will remain separated of the dividers are removed and the system is left to

31 To clarify; the equation states that within a closed system the [instantaneous] rate of change [the differential or 'd'] of entropy [or 'S'] with respect to time is positive. In other words over a period of time a closed system will move towards a state of increased entropy and this will continue without change until a state of thermodynamic equilibrium [or maximum value of S] is reached.

32 Kenneth Sayre, *Cybernetics and the Philosophy of Mind* (London: Routledge and Kegan Paul, 1976), p. 23.

its own devices. In this case, the system 'wants' (if you will) to become entropic. This 'will to disorder' means that the system is predisposed to distribute the elements (in this case the smoke) equally. This means that the eventual mixture of black and white smoke is the most probable state for the system to be in. Likewise, the system which is highly structured according to organisational information, is in a less probable state when being separated into smoke and air. On probability and structural organisation Bertalanffy has stated:

> But entropy, as we have already heard, is a measure of disorder; hence negative entropy or information is a measure of order or of organization since the latter, compared to the distribution at random, is an improbable state.[33]

The Second Law of Thermodynamics has proved highly effective in describing how closed systems behave. However not all systems are closed systems. The limitations the law become evident when one uses it to model observable events which occur within open and dynamic systems. Von Bertalanffy stated:

> Conventional Physics deals only with closed systems, i.e., systems which are considered to be isolated from their environment. Thus physical chemistry tells us about the reactions, their rates, and the chemical equilibria eventually established in a closed vessel where a number of reactants are brought together. Thermodynamics expressly declares that its laws apply only to closed systems. In particular, the second principle of thermodynamics states that in a closed system, a certain quantity called entropy must increase to a maximum and eventually the process comes to stop at a state of equilibrium. [The second principle can be formulated in different ways, one being that entropy is a measure of probability, and so a closed system tends to a state of most probable distribution… So the tendency towards maximum entropy or the most probable distribution is the tendency to maximum disorder.] However, we find systems which by their very nature and definition are not closed systems. Every living organism is essentially an open system. It maintains itself in a continuous inflow and outflow, a building up and breaking down of components, never being, so long as it is alive, in a state of chemical and thermodynamic equilibrium but maintained in a so-called steady state which is distinct from the latter. This is the very essence of that fundamental phenomenon of life which is called metabolism, the chemical process within living cells. What now? Obviously, the conventional formulations of physics are, in principle, inapplicable to

33 Bertalanffy, *General System Theory*, pp. 30–53.

> the living organism qua open system and steady state, and may well suspect that many characteristics of living systems which are paradoxical in view of the laws of physics are a consequence of this fact.[34]

A theory of dynamic systems as is practiced by the systems theoretical approach concerns itself with open systems. In doing so, it can be used to identify a limit to the vocabulary of so-called 'classic' physics, which fails to accommodate the behaviour of open systems. This failure lies in an inability to account for that fact that open systems receive 'neg-entropy' (negative entropy or information) from outside their internal functions. Since neg-entropy effects the internal functioning of these systems they cannot be sufficiently explained according to internal effects alone. The input of external information affects the degree of internal organisation. The system therefore displays less entropy and becomes more ordered and thus it appears that the process towards increasingly probable entropic states suggested by the second law is reversed. Such systems display, for example, evolutionary behaviour (by which a system adapts to its environment), which can be empirically observed but which contradict the world-view of the Second Law of Thermodynamics.

In anticipation of the arguments advanced below a similarity is identified between the methods and limits of 'classic' physics described above (and exemplified by the Second Law of Thermodynamics) and an account of art which also deals with it as if it were a closed system. Such a view is exemplified by Formalist art criticism. Thus when Greenberg famously insisted that; 'The purely plastic or abstract qualities of the art are the only ones that count'[35] he was not only ignoring the institutional systems that served to support the modernist art at the centre of his analysis but was also effectively undermining his own project by failing to acknowledge the function of his own writing in providing regulatory feedback on the growth of American abstract art. Thus for the same reason that the Second Law of Thermodynamics is flawed (namely that it deals with closed, and not

34 Ibid.

35 Clement Greenberg, 'Towards a Newer Laocoon', in Charles Harrison and Paul Wood (eds.), *Art in Theory 1900–1990* (Oxford: Blackwell, 1992), pp. 554–560.

open, systems) so too a formalist analysis of a work of art is limited because it treats the object as a closed system of representation that is not engaged in processes of feedback with the supporting institutional and critical systems by which it is supported and within which it exists. In this book I propose to overcome these limitations by means of a systems theoretical approach that sees the work of art as an open system of representation that itself constitutes a subsystem of the art world which in turn interacts with larger historical, social, philosophical, political and economic systems.

3.3 The whole is greater than the sum of the parts

It was argued above that complexity within systems means that the behaviour of dynamic systems cannot be understood in terms of either the functioning of the elements of the system on their own or the system as isolated from its environment. Further it was argued that a complex system is an organisation of functioning elements with an identity that is recognisable in relation to both those functions and that organisation. This means that in the act of both identifying a complex system and the subsequent analysis of the behaviour of that system the system needs to be considered in its entirety. In other words the whole of the system is a more significant object of critical analysis than its individual components. Further the whole of the system is a different entity than the mere sum of the parts of the system. And the system as a distinct entity can be distinguished from its environment.

Von Bertalanffy locates this move to a holistic approach within the systems theoretical method in relation to a broader epistemological shift away from a method based upon atomism. Von Bertalanffy associates such an atomistic approach with positivism within science and more specifically the approach of his own academic heritage. He writes:

> Having been educated in the philosophical tradition later called the Vienna Circle, I may perhaps hint at the change that has taken place [with regards to the methodological shift of a systems-approach]. The epistemology (and metaphysics) of logical positivism was determined by the ideas of physicalism and re-

> ductionism, and the 'camera theory' of knowledge. These in the view of present day knowledge, are quite simply obsolete.... Compared to the analytical procedure of classical science, with linear causality connecting two variables as a basic category, the investigation of organised wholes of many variables requires new categories of interaction, transaction, teleology, etc. with many problems arising in epistemology, mathematical models and techniques.[36]

Two of the most significant implications of the holist approach are:

a. The system needs to be understood in terms of both its parts and the structure of those parts because everything in a complex system is interconnected.[37]

b. Complex systems operate as a whole; this means that the behaviour, inputs and outputs of such systems are *distributed* over those systems.

Kevin Kelly described the concept of holism in the systems theoretical approach in terms similar to that of von Bertalanffy whilst also describing its interdisciplinary application:

> The spirit of a beehive, the behaviour of an economy, the thinking of a supercomputer, and the life in me are distributed over a multitude of smaller units (which themselves may be distributed). When the sum of the parts can add up to more than the parts, then that extra being (that something from nothing) is distributed among the parts. Whenever we find something from nothing, we find it arising from a field of many interacting smaller pieces. All the mysteries we find most interesting – life, intelligence, evolution – are found in the soil of large distributed systems.[38]

The ramifications of the above are discussed in greater length throughout the rest of this book. In anticipation of these discussions, however, I propose that a holistic critical approach (as opposed to the atomistic viewpoint) which is a key characteristic of systems thinking approach has two significant implications when applied to thinking

36 Bertalanffy, 'Foreword', p. xix. A parallel can arguably be drawn between the Vienna Circle, to which von Bertalanffy alludes, and a particular configuration of scientific art history (also connected with Vienna in the early 20th century).

37 This interconnection takes the form of processes of feedback and distribution as is discussed in more detail throughout the remaining chapters of the book.

38 Kelly, *Out of Control*, p. 469.

about art. These reflect the two implications of the holist approach in general (as given above):

a. In its application within social sciences the holism of the systems theoretical approach necessitates a move from a sociology based on the individual human subject to one based upon the systemic structures from out of which those human subjects emerge as was most famously practiced in Luhmann's sociology based upon his analysis of 'social systems'.[39]

b. In a study of art in terms of systems this attention to the whole system means a shift in attention away from individual works of art onto the structures and systems within which those incidents of art occur. I identify these as the systems of (amongst others): the art gallery; the art market; art discourse; and the artwork itself as a complex system.

3.4 Emergence

The concept of the whole of the system being a more significant conceptual entity than the aggregate of its parts can also be understood in terms of emergence. Emergence occurs when there is sufficient complexity in a system for additional properties or behaviours to emerge from the system. These are often called 'global' properties of the system, as Heylighten observed:

> *Emergence* is a classical concept in systems theory, where it denotes the principle that the global properties defining higher order systems or 'wholes' (e.g. boundaries, organisation, control, ...) can in general not be reduced to the properties of the lower order subsystems or 'parts'. Such irreducible properties are

39 Luhmann's method forms the basis of my own discussions concerning the function of art, the artist and art history within such social systems. See especially Niklas Luhmann, *Social Systems*, trans. John Bednarz and Dirk Baecker (Palo Alto: Stanford University Press, 1995). I do not in this book take up his specific concept of the work of art as a form of communication as is specifically discussed in: Niklas Luhmann, *Art as a Social System*, trans. Eva M. Knodt (Palo Alto: Stanford University Press, 2000).

called emergent. Until now there is no satisfactory theory explaining what characterises emergent properties or what are the conditions for their existence.[40]

The concept of emergence in complex systems has been applied to the self-organisation which is observable in many complex systems (such as, for example the system of an ant colony). Such self-organisation is also observable in the systems of art which appears as an emergent property of complex systems. Emergence has thus been used to explain a wide variety of behaviour observable in complex systems. One of its most significant applications can be found within neuroscience and the discourse of Artificial Intelligence/Life. Here consciousness is understood as an emergent property of a complex cognitive system.[41] As Coveney and Highfield observed:

> Life is also an emergent property, one that arises when physiochemical systems are organised and interact in certain ways. Similarly a human being is an emergent property of huge numbers of cells, a company is more than the sum of its pens, papers, real estate, and personnel, while a city is an emergent property of thousands or millions of human beings. And no one should doubt that out innermost thoughts, our emotions of love and hate, are more than a rush of individual hormones, or the firing of individual neurons in the brain. The study of complexity, through its emphasis on emergent properties, goes some way to restoring a balance between the spiritual and materialistic sides of our nature.[42]

40 Francis Heylighen, 'Self-Organization, Emergence and the Architecture of Complexity', *Proceedings of the 1st European Conference on System Science* (AFCET, 1989), pp. 23–32.

41 The literature on this area is rapidly expanding. An early philosophical study concerning a systems-thinking application to the mind/body debate in materiallist terms is found in Kenneth Sayre, *Cybernetics and the Philosophy of Mind* (London: Routledge and Kegan Paul, 1976). However, Sayre's arguments were not testable without the sophisticated computer technology that has emerged from the late 1980s onwards. Excellent overviews of the major current philosophical debates are found in: Terrance Horgan and John Tienson (eds.), *Connectionism and the Philosophy of Mind* (Dordrecht: Kluwer, 1991); W. Ramsey et al. (eds.), *Philosophy and Connectionist Theory* (Hove and London: Lawrence Erlbaum Associates, 1991); Margaret Boden (ed.), *The Philosophy of Artificial Life* (Oxford: Oxford University Press, 1996) and *The Philosophy of Artificial Intelligence* (Oxford: Oxford University Press, 1990).

42 Coveney/Highfield, *Frontiers of Complexity*, p. 330.

I argue that the very concept of art is itself an emergent property of the complex systems with which it is associated. These include the systems of the art gallery, the art market, art discourse and the complex networks of artworks themselves.

3.5 Complexity

Complexity is the most important concept of the types of systems which I discuss throughout this book. It is also, however, the most difficult to define precisely. Cilliers noted: 'The concept remains elusive at both the qualitative and quantitative levels.'[43] Likewise Coveney and Highfield have also written about the difficulty in defining the concept: 'This word is often used in an infuriatingly vague sense by scientists, who may mean different things by it.'[44] Likewise, Heylighten gave the following commentary on the difficulty in providing a specific definition of complexity:

> Still, I believe there is a common, 'objective' core in the different concepts of complexity. Let us go back to the original Latin word 'complexus', which means 'entwined', 'twisted together'. This may be interpreted in the following way: in order to have a complex you need two or more components, which are joined in such a way that it is difficult to separate them. Similarly, the Oxford Dictionary defines something as 'complex' if it is 'made of (usually several) closely connected parts'. Here we find the basic duality between parts which are at the same time distinct and connected. Intuitively then, a system would be more complex if more parts could be distinguished, and if more connections between them existed. More parts to be represented means more extensive models, which require more time to be searched or computed. Since the components of a complex cannot be separated without destroying it, the method of analysis or decomposition into independent modules cannot be used to develop or simplify such models. This implies that complex entities will be difficult to model, that eventual models will be difficult to use for prediction or control,

43 Paul Cilliers, *Complexity and Postmodernity* (London: Routledge, 1998), p. 2.
44 Coveney/Highfield, *Frontiers of Complexity*, p. 14.

and that problems will be difficult to solve. This accounts for the connotation of difficult, which the word 'complex' has received in later periods.[45]

From the perspective of systems thinking, when it is said that systems are 'complex' something quite specific is meant.

First, complex systems are open systems. This means that they receive organising information from outside their boundaries which increases the order within the system.

Second, it does not just mean that a system is very big, nor even does it mean that it is complicated. Instead 'complexity' refers to the way in which the system functions in a dynamic manner. Complexity emerges from dynamic as opposed to simple systems. This means that the way in which a complex system functions is not explained through an explanation of how every bit of the system works; instead something of the system as a functioning whole is required. As Cilliers observed, 'there is neither something at the level below (a source) nor at a level above (a meta-description), capable of capturing the essence of complexity.'[46] Further a complete description of a complex system must be as complex as the system that it is describing. This is because if the description is simpler than what it is describing then the complexity will have been misrepresented.

Nonetheless the following characteristics of complex systems can be identified in relation to the following concepts: a. complex (as opposed to complicated); b. dynamic; c. interconnected; d. whole (as opposed to a aggregate of its parts); e. non-linear; f. balance between connection and differentiation. I will now discuss each of these concepts in turn.

3.5.1 Complex, not Complicated

When a system is identified as complex it is thereby differentiated from a system which is merely complicated. To call something 'complicated' is to say that it can be recognised as being composed of

45 Francis Heylighten, 'What is complexity?' (Dec 9, 1996), published online at the *Principia Cybernetica Web* http://pespmc1.vub.ac.be/COMPLEXI.html (29/09/07).

46 Cilliers, *Complexity and Postmodernity*, p. 3.

many different parts. These parts could be organised in simple manner but be complicated by virtue of the sheer number of parts involved. An example of this would be an imaginary jigsaw puzzle of 100,000 pieces. The puzzle is complicated because of all its many parts. And it would require this complication to be negotiated in order for the puzzle to be completed. However despite this numerical complication the way in which a jigsaw works is, in actuality, very simple. It is not complex. Likewise an organisation could be made of many parts and be structured in a complicated way but still not be complex because it does not function in a complex way. As Cilliers has observed: 'I have heard it said (by someone from France, of course) that a jumbo jet is complicated, but that a mayonnaise is complex.'[47]

3.5.2 Dynamic

Complexity emerges from dynamic systems. There are three aspects of a system which allow it to function dynamically; and hence in a complex manner. These are:

a. The type of elements within a system. They must be those which will interact with one another in different ways. This means that each element is capable of behaving in different ways depending upon the other elements that it is interacting with and the mode of that interaction. In short each part of a dynamic system must be flexible enough to work with other parts of the system in creating the dynamic whole that is a complex system.

b. The way in which those elements are organised. The elements of a complex system must be organised in such a way so as to allow for a dynamic interaction between the (adaptable) elements.

c. The manner in which the elements interact with one another. Self-evidently, in order to be complex a system must actualise its potential for dynamic interaction. That is it must display complex behaviour.

47 Ibid.

3.5.3 Interconnected

One of the reasons that certain systems are dynamic and complex is because of the high level of interaction amongst the component parts. For example a complex system such as a neural network operates dynamically by virtue of a high level of interconnection between the component parts. This serves to illustrate that what makes an arrangement dynamic (and hence complex) is the different ways in which the components of a system can interact. Within a jumbo jet there is a specific way in which the parts can interact. This interaction is set within the parameters of the jumbo jet's function and there is little potential allowed for deviation from this function. For example the tyres of the plane have a specific function and cannot interact with the air-conditioning system. If the parts were arranged in a different order then the object wouldn't be a jumbo jet. There are clearly a great many parts to the jet and they perform a great many functions – I couldn't build one in my garage – but this is just an expression of the complication of the jet's design rather than its complexity. A mayonnaise on the other hand is complex because whilst each element of the emulsion (egg yolk, oil, acid and salt) contributes to the overall system the mayonnaise as a whole and they are in a symbiotic (and non-linear) relationship with one another.

3.5.4 Complex Holism

An understanding of a complex system must operate at the level of the system as a whole and not on an atomistic level which reduces the system to its component parts.

To be complex, a system must be both numerous and diverse. As Coveney and Highfield have stated, 'the complexity of a problem is defined [in mathematical terms as] the number of mathematical operations needed to solve it.'[48] Thus complex systems are similar to complicated systems in so far as both are composed of a large number of component parts. However, in complex systems this numerousness creates a diverse and complex interaction between those elements. For

48 Coveney/Highfield, *Frontiers of Complexity*, p. 14.

example a collection of five ants is not complex yet an ant colony is. This is because the potential for dynamic interaction between the five ants is limited (by the number of ants) whereas the potential for dynamic and complex interaction between the ants in the colony is enormous. A small system is easy to describe by describing the individual elements and their behaviours (through differential equations for example). However in a large and complex system this description of the elements alone is not sufficient to describe and understand the way in which the system works. Thus, in order to account for a complex system, the level of description needs to be at the system as a whole.

3.5.5 Non-linearity

If interactions between elements within a dynamic system are both diverse and non-linear then they will produce complex behaviour; and further they will contribute to the complexity of the system. By diverse interactions it is meant that elements within the system will interact with many other elements of the system; and do so in varied ways. For example ants in a colony can interact with one another in a variety of ways that would not be possible for the elements in a jumbo jet to do. By non-linear it is meant that interactions need not happen in the same order for subsequent interactions. Non-linearity is a defining feature of complex systems and also allows for another key characteristic of complex systems namely positive feedback within complex systems (as is discussed in more detail below) where small causes can have large effects.

The non-linearity of dynamic/complex systems has two other implications.

a. Dynamic systems are not in a state of equilibrium. They are constantly changing their state and do not remain the same over time (that is that their structure, rather than just their physical constitution, and operations adapt and evolve); hence their dynamism. As Cilliers observed:

> Complex systems operate under conditions far from equilibrium. There has to be a constant flow of energy to maintain the organisation of the system and to ensure its survival. Equilibrium is another word for death.[49]

b. A dynamic system develops in a non-linear fashion because it has 'memory'. This is to say that previous operations within the system are 'remembered' by the system. The weather system 'remembers' the butterflies wings flapping when the tidal wave breaks. This argument is discussed in greater length in my discussion below on 'lock-in' and systemic memory.

3.5.6 The Balance between connection and differentiation

The emergence of complexity depends upon two aspects of a system: a. connection and b. differentiation. Both of these two elements need to be balanced in order for complexity to emerge. Connection means that parts of the system need to be interconnected. The greater the level of interconnection the greater the level of order within the system. This is because more information will be required to describe the system; in other words a connected system will require more data to model what goes on within the system. The connected system will thus display less entropy (or disorder) and more neg-entropy than an unconnected one. Complete connection within a system (in which *every* element is connected with every other element) leads to a completely ordered system. An example of a completely order system would be a perfect crystal in which the position of each molecule within the crystal is related to every other molecule in a completely interconnected structure. Differentiation in a system, on the other hand, means that the components of the system are distinct from one another. This means that there will be variety in the system and that different parts of the system will behave differently. Increased differentiation within a system contributes not only to the complexity of a system but also, contrary to connection, to its disorder. Complete differentiation within a system leads to complete disorder. An example of such disorder would be a gas in which the movement of the molecules within the gas in its environment are independent of one another. A system can

49 Cilliers, *Complexity and Postmodernity*, p. 4.

only be complex if a balance between connection and differentiation is present. This means that a complex system is both ordered and disordered. It has a structure but also variety and capacity for change and dynamic behaviour. An increase in both connection and differentiation within a system can contribute to that system's complexity this is because with more structural variety more dynamic behaviour, for different types of operations and thus change is possible. This is not the case in a totally connected system (such as a crystal) which is too ordered to allow the possibility of dynamic change or a totally differentiated system which will also not change because it has reached a state of maximum entropy. A complex system is therefore one balanced between order and disorder. This balance allows the system to be creative and to change over time under its own impetus (what Varela called auto-poiesis)[50] but to do so in a way that means that it retains its own recognisable identity. This complex balance between chaos and order has lead to complexity being called *The Edge of Chaos*. Such systems on the 'Edge of Chaos' can adapt and evolve as Goodwin argued:

> [The edge of chaos] is almost a theorem about life, the universe and everything that is complex and non-linear (which is *nearly* everything.) Speaking anthropomorphically the edge of chaos is a good place to be in a constantly changing world because from there you can always explore the patterns of order that are available and try them out for their appropriateness to the current situation [...]. Complex systems that can evolve will always be near the edge of chaos, poised for the creative step into emergent novelty that is the essence of the evolutionary process.[51]

50 See Francisco Varela, *Autopoiesis: A Theory of Living Organisation* (Amsterdam: North Holland, 1981).

51 Brian Goodwin, *How the Leopard Changed its Spots* (London: Weidenfield and Nicolson, 1994), p. 175. Kaufman also argued that the edge of chaos is a 'powerful new framework to understand evolutionary biology', which can be used to explain the evolution of biological systems. See Stuart Kauffman, *The Origins of Order* (Oxford: Oxford University Press, 1993), p. 26.

3.6 Isomorphism

Isomorphism the final key concept of a systems theoretical approach I explain here is, arguably, the most important for method; that is my own application of dynamical systems theory to thinking about art. This is due to that fact that because dynamical systems theory proposes that isomorphisms (or structural similarities) are identifiable amongst diverse systems it can be legitimately applied to other discourses including the systems of art. When general system theory was originally formulated by von Bertalanffy, it was as a solution to what he recognised as the failure of the increasingly isolated and specialised sciences[52] to accommodate certain phenomena. Hence the failed application of the second law of thermodynamics to open systems as outlined above. Thus the aim of his method was to identify structural similarities or *isomorphic tendencies* having currency in the gaps between isolated scientific disciplines such as biology, chemistry and physics. The goal was to find these more general principles and thus save the language of the specialist from unintelligibility, metaphysics and isolation from the world which I discussed at the beginning of the chapter.

Dynamical systems theory is thus characterised by an interdisciplinary method seeking to identify isomorphisms between different systems and works on the basic principle that there are universal principles according to which all dynamic systems, no matter what form they take, are organised. As von Bertalanffy observed:

> Speaking in 'material' language, it means that the world, i.e. the total of observable events, shows structural uniformities, manifesting themselves by isomorphic traces of order in the different levels or realms.[53]

52 This can be compared with both Husserl's comments on the 'crisis' of the sciences in the modern age: Edmund Husserl, *The Crisis of European Sciences and Transcendental Phenomenology* (Evanston: Northwestern University Press, 1970); and Lyotard's arguments in *The Postmodern Condition* (Manchester: Manchester University Press, 1976). Both argue that the crisis of the condition of knowledge in modernity is defined by an increasing specialisation of disciplines which facilitates their increasing autonomy and isolation.

53 Bertalanffy, *General System Theory*, p. 49.

Thus the universal concept of the 'system' serves as the overarching conceptual framework by which a multitude of phenomena from diverse fields of study can be explained. In the spirit of such an interdisciplinary method further general and 'isomorphic' systems principles can be observed in the systems of art.

4. Applications

Given the suggestiveness of the systems theoretical approach it has been taken up in a number of practices. Management studies for example has found the concept of 'system' as a highly useful concept in understanding corporate and management structures and the literature on 'Management Systems' and so forth is vast. Arguably the most productive recent application has been the application of systems thinking and complexity to the questions and philosophy of Artificial Intelligence and Artificial Life. In artistic practice there has been some application of 'systems' to inform artistic practice as I discuss in further detail in Chapter 3 on the systems aesthetic. This is most notable in the work of Jack Burnham, who first invoked systems thinking in the 1960s; and Hans Haacke's numerous artistic engagements with artistic 'systems' (illus. 1). More recently the *Biotica* project at the RCA (1999) attempted an artistic synthesis of modelling of Artificial Life and installation art.[54] To my knowledge there has been no attempt systematically to apply the systems theoretical approach to art history in the English language (in book length form at least).[55]

54 See Richard Brown, *Biotica: Art Emergence and Artificial Life* (London: RCA, 2001).

55 Notable articles include E. Sevänen, 'Art as an Autopoietic Sub-System of Modern Society', *Theory, Culture and Society*, vol. 18 (1) (London: Sage, 2001) and F.E. Rakuschan, 'Art and Repugnance: Form as Anti-Form, Art as Anti-Art, Market as Anti-Market', was published on the website of the *European Institute for Progressive Cultural Policies*, http://www.eipcp.net/diskurs/d05/text/rakuschan_kunst_en.html (15/10/2007). Kitty Zijlmans has made a valuable contribution to the field. See for example her 'Kunstgeschichte als

However, I identify two precedents for my arguments and method below.

4.1 Luhmann – Social Systems and Art as a Social System

The work of the German sociologist Niklas Luhmann is exemplary of the usefulness of the systems theoretical approach when applied to the humanities.[56]

In German speaking scholarship Niklas Luhmann is widely regarded as amongst the most important and widely read sociologists of the late 20th century. That this degree of recognition of his theory of social systems was not, by the time of his death in 1998, reflected in Anglophone studies in general, and to studies on art in particular, is a situation which is slowly being rectified with the ongoing translation of his vast, diverse and highly sophisticated work. His general theory is outlined in *Social Systems* and numerous other books including studies on the systems of law, economics and science. His main theories on art are outlined in *Art as a Social System* in which he explored his anti-ontological description of art. Art is, for Luhmann, constituted by the system of art, itself a sub-system of society.

Society, Luhmann argues, is comprised of a variety of subsystems. These are operatively closed and functionally distinct from one another. Each system operates according to its own internal and self-defined rules. These systems include economic, political, legal, scientific, religious and educational systems. Like these, the art system functions according to an intrinsic set of processes. However, despite the local differences between these particular systems they, nonetheless, exhibit what Luhmann calls 'comparable structures'. They are thus receptive to the comparative analysis of systems theory.

Systemtheorie', *Gesichtspunkte. Kunstgeschichte heute* (Berlin: D. Reimer Verlag, 1995), pp. 251–277.

56 However, note that I do not attempt an evaluation of Luhmann's specific approach to art (as it is discussed in *Art as a Social System*) within this book. Instead I discuss Luhmann's work as only one example of the application of systems thinking to questions of cultural analysis.

Luhmann's analysis of social systems operates according to the following key terms:

a. *Systems of observation*. These are systems whose defining function it is to observe. For example the economic system is a system which observes economic value.

b. *Systems of differentiation*. These are systems whose defining function is to differentiate (between systems, and, crucially, between systems and their environment).

c. *Systems of Communication*. These are systems whose defining function is to operate in communication (transfer of information as it is understood by Shannon's conception of information theory). [57]

Luhmann claimed that his project should be understood as a 're-writing [of] the theory of society' through the systems theoretical approach. This, he argued, implied a paradigm shift:

> At the structural level, from stratification to functional differentiation. The unity of society is not to be sought in ethico-political demands, but rather in the emergence of comparable conditions in systems as diverse as religion or the monetary economy, science or art, intimate relationships or politics – despite extreme differences between the functional and operational modes of these systems. Our theoretical proposition offers the following: a clear demarcation of external system boundaries of different domains and comparability between different systems.[58]

What Luhmann proposed was thus nothing less than a move, in sociological practice, away from an analysis of the structural 'stratification' of society in terms of critiques of societal organisations and hierarchies and social relationships[59] to a sociological analysis in

57 Luhmann's adaptation of Shannon's information theory to the specific issue of communication within social systems is introduced in the collection of essays: Niklas Luhmann, *Theories of Distinction*, ed. by William Rasch (Stanford: Stanford University Press, 2002), esp. Part III: 'What is Communication?' For Shannon's own definition of Information Theory see Shannon/Weaver, *The Mathematical Theory of Communication*.

58 Luhmann, *Art as a Social System*, p.2.

59 Such a method was exemplified by the earlier attempt to 'systematise' sociology practiced by Talcott Parsons. See Talcott Parsons, *The Social System* (Glencoe: Free Press, 1951), for an outline of his largely structuralist approach, which has its roots in the social structuralism of Weber and Levi-Strauss. This

terms continuous with the strategies of systems theory. One of Luhmann's translators, Eva M Knodt, elaborated on this paradigm shift in the following manner:

> General systems theory is the result of two subsequent paradigm shifts, which moved the problem of order from the fringes of metaphysical speculation to the centre of scientific research. In the first of these shifts, initiated by the German biophysiologist Ludwig von Bertalanffy in the mid-1950s, the metaphysical distinction between part and whole was replaced by the distinction between system and environment… In a second shift, the system/environment distinction was redefined within a general theory of self-referential systems.[60]

Hence in Luhmann's work there is a move away from an analysis of social structures *in themselves*, which in structuralist sociology are pre-empted before they are deconstructed, to analyses of the ways in which these structures emerge from the complex matrices of communication which occur at the level of the social system.[61]

For Luhmann, the social systems of his analyses are not composed as rigid macro-structures which dictate behaviour from the top

mode of sociology requires an understanding of society in terms of social structure. Luhmann's contribution was his recognition that the structure he observed was a function of, and thus relative to, his own discursive system rather than an extra-discursive 'object'.

60 Eva M. Knodt, 'Foreword', in Luhmann, *Social Systems*, p. xxi.

61 This move away from an analysis of the structure of social systems toward examinations of the functional patterns of social behaviours (in terms of the dynamics of complex systems) is characteristic of a more general skepticism toward structural analytical strategies motivated by a rejection of modernism. This rejection of modernism forms part of larger debates, importantly the debate between Luhmann and Habermas. For more on this see William Rasch, *Niklas Luhmann's Modernity* (Stanford: Stanford University Press, 2000). See also W. Rasch and Cary Wolfe (eds.), *Observing Complexity: Systems Theory and Postmodernity* (Minneapolis: University of Minnesota Press, 2000), which contains Luhmann's own paper, 'Why Does Society Describe Itself as Postmodern?' alongside commentaries on the debate. A good English introduction to Habermas' position is found in: Jürgen Habermas, *The Philosophical Discourse of Modernity: 12 Lectures* (Cambridge, MA: MIT Press, 1987); Habermas, 'Modernity versus Postmodernity', trans. Seyla Benhabib, *New German Critique*, 22 (1981), pp. 3–14 and Habermas, 'Modernity – An Incomplete Project', in Hal Foster (ed.), *The Anti-Aesthetic* (Port Townsend: The Bay Press, 1983), pp. 3–15.

down by organising it and systematising it. In other words they are not a scaffolding which supports social activity or bureaucratically orders social relationships and communications. Instead Luhmann suggests that social systems are formed from the bottom up; that is that they are emergent properties of the functioning of social relations on the micro (not macro) level. Social systems are, therefore, a collection of functions, or to use Luhmann's terms, a collection of 'operations':

> Operations (conscious perceptions as well as communications) are nothing more than events. They cannot persist, nor can they be altered. They emerge and vanish in the same instant, taking no more time than is needed to fulfil the function of an element that cannot be decomposed any further.[62]

When this thinking is applied to systems of art it produces perhaps Luhmann's most radical insight on art; this is his anti-ontological definition. His definition is independent of particular genres such as painting or music. Instead it is a non media-specific theory of 'the medium of art' that produces a special type of communication within the system of art. This involves an inversion of a traditionally conceived relationship between medium and form. The traditional notion that 'medium constitutes form' is replaced by the notion that 'form constitutes medium'. The example Luhmann gives (in 'The Medium of Art' (1990)) is music. Music is heard (observed) as music only by those 'who can also hear the uncoupled space in which the music plays'. In other words 'whereas we normally hear noises as differences to silence and are thereby made attentive, music presupposes this attention and compels it to the observation of a 2nd difference – that between medium and form.' We may 'observe' the medium of music, because we have paid attention to the sensuous form of the music's manifestation. In this sense, then, *form constitutes medium*.

4.1.1 Psychic Systems: Luhmann and the Human Subject

Luhmann conceives of individual humans within social systems in terms of what he defined as 'psychic systems'. It is this identification of the individual in the same terms as society (i.e. as systems) that

62 Luhmann, *Art as a Social System*, p. 49.

allows him to discuss operational isomorphisms shared between them. His discussions on psychic systems identify the following key points:

a. Consciousness is a (closed) psychic system (differentiated from its environment and other psychic systems);

b. Psychic systems are characterised by the operations of perceptions particular to individual psychic systems. That is that self-consciousness is a particular type of perception; that this perception is one of self-recognition/apperception and hence can be identified, in systems theoretical terms as self-reflexive or cybernetic;

c. Psychic systems operate within larger (social) systems;

d. The operations of social systems have effects on the operations of psychic systems (and vice versa) and therefore that:

e. Perceptions within psychic systems (the operations of psychic systems) are involved in relations of feedback, exchange and control with social systems.

Hence Luhmann argued that conditions of social systems will effect the operative conditions of the psychic systems which operate within them.[63] He further argued that within social systems inter-

63 A particular aspect of this relationship between psychic and social systems is the interdependence between perceptions or outputs of psychic systems and the overall technological state of the social system. Luhmann articulated this interdependence in terms of the 'meaning' produced by the organisational information and the output of both psychic and social systems: 'Psychic and Social Systems have evolved together. At any time the one kind of system is the necessary environment of the other. This necessity is grounded in the evolution that makes these kinds of systems possible. Persons cannot emerge and continue to exist without social systems, nor can social systems without persons. This co-evolution has led to a common achievement, employed by psychic as well as social systems. Both kinds of systems are ordered according to it, and for both it is binding as the indispensable, undeniable for of their complexity and self-reference. We call this evolutionary achievement "meaning"' (Luhmann, *Social Systems*, p. 59). He also claims that: 'Both kinds of systems emerge by the path of co-evolution. One is impossible without the other, and vice-versa. They must, so to speak, differentiate themselves in respect to meaning. Meaning is the true "substance" of this emergent evolutionary level. It is therefore false (or, more gently, it is a falsely chosen anthropocentrism) to assign the psychic, that is, the conscious, anchorage a sort of ontological priority over the social. It is impossible to find a "supporting substance" for meaning. Meaning supports itself in that it enables its own self-referential repro-

actions take place between psychic systems. However Luhmann argues that this interaction between psychic systems does not occur at the level of 'communication' within social systems but rather that it takes place in a manner different to the discreet transfer of information between minds. This is because, Luhmann argues, psychic systems are closed to both themselves and one another; they have 'operative closure' which means that they have clearly delineated boundaries within which they operate. Luhmann argues that because of this operative closure perception itself (which is the operations of a closed psychic system in terms of that system's self-recognition, or self awareness) cannot be directly communicated to another psychic system. In *Art as a Social System* Luhmann stated:

> One thing is certain; if it is generally true that psychic operations, not to speak of those of the living system, can never be executed in another consciousness, which because of its complexity and historically self-referential mode of operation remains opaque, then this holds also for the artist distanced by his work and for his admirers as well – no more and no less, for inaccessibility does not allow for augmentation.[64]

In *Art as a Social System* Luhmann argues that art has a special function within social systems because it allows for the replication and hence communication of the perceptions and operations of these closed psychic systems between one another. However the arguments outlined at greater length in this book as a whole will not investigate the problematic issue of intersubjectivity between psychic systems and the nature that art may play in this. I will instead use Luhmann's arguments concerning the discursive self-reflexivity of systems theory as a means for art discourse to reflect upon its operations through the adoption of a systems theoretical approach.

4.1.2 Luhmann and Complexity

Another key aspect of Luhmann's contribution to sociology was his use of the principle of complexity (and complex systems) as a descrip-

duction. And only the forms of this reproduction differentiate psychic and social structures' (Luhmann, *Social Systems*, pp. 97 f.).

64 Luhmann, *Art as a Social System*, p. 13.

tive organising principle of social systems. Thus both complexity and complex systems are used in the analysis of social systems.

Luhmann engages with complexity in two fundamental ways.

a. The particular behavioural patterns of the social systems which he described are specific to complex systems; that is they are dynamic, unpredictable, self-organising (auto-catalytic) and display emergence.

b. As a consequence of ensuring that his own discursive system (such as that presented in the 600 pages of *Social Systems*) satisfactorily accounts for the complexity in the social systems he observed, his own textual and theoretical system is mimetic of the complexity he discusses. This means that texts such as *Social Systems* are themselves complex. This is something Luhmann himself recognised at different stages in the reception of *Social Systems*. In the original preface to *Social Systems* he wrote:

> A sociological theory that wants to consolidate the conditions of the discipline must not only be more complex, it must be much more complex that the classical authors and their interpreters – even Parsons – had thought. This requires different theoretical precautions in regard to validity and connectivity, internally as well as externally, and it requires, not least, building the reflection of complexity (and the concept of complexity) into the theory itself.[65]

And then again in 1991 in the preface to the English edition of *Social Systems* he wrote:

> This is not an easy book. It does not accommodate those who prefer a quick and easy read, yet do not want to die without a taste of systems theory. This holds for the German text, too. If one seriously undertakes to work out a comprehensive theory of the social and strives for sufficient conceptual precision, abstracttion and complexity in the conceptual architecture are unavoidable.[66]

4.1.3 Luhmann and Self-Reflexivity

Luhmann claimed that the synthesis of a systems theoretical approach and sociology radicalises the methods of sociology by allowing for a methodological self-reflexivity which he identifies as lacking. In short,

65 Luhmann, 'Instead of a Preface to the English Edition: On the Concepts "Subject" and "Action"', *Social Systems*, p. xlix.

66 Ibid., p. xxxvii.

Luhmann used systems theory as a means by which to achieve a theory which could include within its own theoretical framework a reflection upon itself *on its own terms*. Thus through the use of systems theory as a strategy of theoretical self-recognition Luhmann was able to claim a generality for his theoretical position. This was because, he claimed, the systems theoretical approach:

> [...] deals with everything social and not just sections (as, for example, strata and mobility, particularities of modern society and patterns of interaction, etc.)
>
> Theories that claim universality are easily recognised by the fact that they appear as their own object. (If they wanted to exclude themselves, they would have to surrender the claim to universality.) Thus it is – and this holds for all 'global theories' (including e.g., quantum physics) – that specific areas of the classical theory of science are suspended, in particular, everything having to do with independent confirmation of the theory's claim to truth [...].
>
> Therefore, theories that make a claim to universality are self-referential. At the same time, they always learn something about themselves from their objects. Therefore they are forced, as if by their own logic, to accept a limitation of their meaning.[67]

It is my intention throughout this book to use a similar approach. In doing so, I employ the systems theoretical approach in a two-fold process of critical reflection. Firstly I use it to reflect upon the systems of art; and secondly to *reflect upon that reflection* as it is itself situated within a discursive system with its own aims, methods and outcomes.

4.2 Cilliers – Postmodernity and Dynamical Systems Theory

Paul Cilliers has provided an excellent commentary on the observable connections between, what he identifies as, *Complexity and Postmodernism* (1998). Cilliers suggests that a critical impetus can be identified between the critique of the metaphysical systems of structuralism (be they social, political, moral or philosophical) that informs postmodernity and the multiple and relative perspectives of systems theory.

67 Ibid., p. xlix.

In his discussion Cilliers argued that the studies of complexity and postmodernity (understood as deconstructionism) operate according to a shared set of values. He stated:

> Saussure's 'structural' model of language remains a landmark in the study of complex systems. His primary insight – that meaning is generated through a system of differences – remains an excellent way of conceptualising the relationships in a complex system. His model is somewhat 'rigid', but Derrida's transformation of the system of the system by means of a sophisticated description of how the relationships interact in time (using the notion of *différance*) provides us with an excellent way of conceptualising the dynamics of complex systems from a philosophical perspective.

And:

> Since it is based on a system of relationships, the post-structural inquiry into the nature of language helps us to therorise about the dynamics of the interaction in complex systems. In other words, the dynamics that generates meaning in language can be used to describe the dynamics of complex systems in general.[68]

From a systems theoretical approach Cilliers identified language, as Saussure and Barthes[69] et al. do, as a (normative) system of differentiation in which the sign is a 'node in a network of relationships'. Language is hence seen as a *system* of representation. He concluded that: 'Put in the language of systems theory, Saussure still understands language as a closed system, whereas Derrida wants to argue for language as an open system [f.n. p. 43][70] and in which, meaning is determined by the dynamic relationships between components of the system [f.n. p. 46].'[71]

68 Cilliers, *Complexity and Postmodernism*, p. 37. A central theme of Cilliers' arguments concern connectionism and representation within complex systems in relation to a conception of postmodernity.

69 'Semiology aims to take in any system of signs, whatever their substance and limits; images and gestures, musical sounds, objects and the complex association of all of these, which form the content of ritual, convention or public entertainment: These constitute if not languages, at least systems of signification.' Roland Barthes, *Elements of Semiology*, trans. Annette Lavers and Colin Smith (London: Jonathan Cape, 1967), p. 9.

70 Cilliers, *Complexity and Postmodernism*, p.42.

71 Ibid., p. 46.

Through his example of linguistic analysis Cilliers provides one case of the possible interconnections between the contemporary epistemological technologies of post-structuralism and the systems theoretical approach. He does this through identifying a correspondence between the post-structural critique of language, in which meaning is understood as an unstable, complex, chaotic, non-linear system and the systems thinking view of unstable, complex, chaotic, non-linear systems in general.[72]

The correspondence between theories of complexity and post-modernity can be understood in terms of a definition of modernity. This is to say that a common ethos is shared between the two can be couched in terms of the project of (social and philosophical) modernity; and more specifically modernity as understood as the proliferation of the establishment of *systems* (of meaning, differentiation, control.) As Rasch observed; modernity, postmodernity and systems thinking can be conceived of as conterminous.

> If the initial, eighteenth-century differentiation of reason associated with the name of Kant came to be seen as the problem of modernity that supposedly needed overcoming, then the subsequent uncontrolled proliferation of languages, system rationalities, and observer positions would seem to signal unsurpassable modernity's irreversible triumph. [...] Thus, if the critique itself is legitimate, then it must be located within the heart of modernity itself. Postmodernism's skeptical, ironic, and unrelenting critique of modernity's more ambitious projects must, in fact, be the quintessential modern gesture, a type of immune system attacking the anxieties that have produced the fevered delusions of foundation and future reconciliation [...]. Postmodernity, were one to retain the word, would then be seen not as the result of the exhaustion or of the dis-

72 This is discussed further in Katherine Hayles (ed.), *Chaos and Order* (Chicago: University of Chicago Press, 1991), which contains several discussions on the application of complex dynamics to a study of literature; Hayles, *How We Became Posthuman: Virtual Bodies in Cybernetics, Literature, and Informatics* (Chicago: University of Chicago Press, 1999), where Hayles further applies systems-thinking and in particular artificial intelligence (connectionism) to a study of literature. She had already visited this theme in Hayles, *Chaos Bound: Orderly Disorder in Contemporary Literature and Science* (Ithaca: Cornell University Press, 1993). Also Espen Aarseth, *Cybertext: Perspectives on Ergodic Literature* (Baltimore: Johns Hopkins University Press, 1997), is a thorough and thought provoking study of hypertext and its history as a literary form in relation to a systems-thinking approach (broadly conceived).

illusionment with modernity but rather as a new realization of modernity's self-imposed limits.[73]

Rasch's observations were being used specifically to identify Luhmann's systems theoretical approach as a continuation of, rather than a break from, modernity's project of self-criticism and self-determination.[74] However, dynamical systems theory as a whole can be similarly characterised. In doing so dynamical systems thinking can be re-configured as a *post-metaphysical modernism*; one in which the 'deep-structure' of systems is identified but with two further implications which demonstrate its affinity with postmodernity. Within the systems-theoretical approach, systems are identified:

a. in order that their structures and processes by which 'meaning' is produced can be *deconstructed*; hence identifying dynamical systems-theory with deconstructionism in general;

b. as being relative to the discursive systems by which they are identified and represented; hence identifying dynamical systems theory with the postmodern 'incredulity toward metanarratives'.

This isomorphism shared between modernity, postmodernity and dynamical systems theory can also be compared to what Rasch identified in Luhmann's praxis as:

> The ability to perceive this deep structure [which] becomes the ability to read surface complexities symptomatically as manifestations of a rationally comprehensible totality [...]. For theory to have social import – which is to say for theory to live up to its purported imperative to change and not just understand the world – it must be able to articulate the connection between deep and surface structure in such a way as to make the deep structure eminently alterable. It must grasp the deep structure, in other words, *from some perspective that is itself not determined by that structure.* To fail to do so would be to fail in one's duty as either a Marxian historical materialist or a Straussian political philosopher.[75]

It is proposed that the 'deep structure' of which Rasch talks can be articulated in terms of the differential *system* of contingencies, but

73 Rasch, *Niklas Luhmann's Modernity*, p. 9.

74 For Luhmann's own commentary on this see his 'Deconstruction as Second-Order Observing', in *New Literary History*, 24, 4 (1993), pp. 763–782.

75 Rasch, *Niklas Luhmann's Modernity*, p. 6.

which is nonetheless a function of its own discursive system and is not an extra-discursive structure. This allows the representation of (Marxian) base and superstructure (of communicative technology and the social system) as inextricably linked and not as a dichotomy of conflicting forces but rather as an organic and systemic unity which displays the behavioural patterns of a complex system such as internal self-organisation, auto-catalytic behaviour and emergent properties.

Rasch continued:

> If the conflict of perspectives is not to be reconciled from a higher-order meta-perspective; if, in other words, a universal perspective of the morally Good or ontologically True cannot be occupied; if, rather, such perspective must compete on the same level as all others, with no hope of logical or divine resolution; then, ironically, the statement that describes this state of affairs must 'pose' as a metastatement and 'occupy' this impossible metaperspective. To describe adequately the ineluctable necessity of the contingency of modernity, the perspective of the metaperspective has to be filled with the assertion that there is no metaperspective, and, thus, the description of modernity as contingent has to serve as modernity's 'transcendental' ground.[76]

It is from a new, *multiple and contingent* meta-perspective of which Rasch talks that dynamical systems theory emerges as an epistemological, heuristic system of observation which is isomorphic to postmodernism. Furthermore, as is explored in the following chapters, it can be used as a strategy to think about the systems of art.

76 Ibid., pp. 24 ff. One of Rasch's arguments is that Luhmann tackles the problem of the meta-perspective of discourse from within the contingencies of the modern perspective through his application of the paradigms of systems thinking.

Chapter 2: The Artworld as a Complex System – Complex and Distributed Representational Systems and the Artworld

> There's a machine which is called History of Art, which is a structure. And the artist [sic] fits in this only because he or she is needed for this structure. If, for example, the History of Art needs some parallel lines, there is an individual who makes parallel lines. And this individual fits into this machine, which works by itself; it doesn't care about people or anything else, it just goes by itself.
>
> Alexander Melamid[1]

This chapter is about how art is represented in the system of the artworld. I will discuss the following three issues:

1. A reformulation of the institutional theory of art from a systems theoretical perspective. I will show that Danto's concept of the artworld can be reconceived in systems theoretical terms as a system of art. Observed as a system, the artworld can be seen to contain the following complex systems: a. the work of art as a complex system; b. the art gallery and market as a complex system; c. art discourse as a complex system. Art, I argue, emerges from these complex systems of representation.

2. An elaboration of the concepts of 'positive feedback' and 'lock-in' as key characteristics of complex systems in their application to a systems-theoretical account of the institutional theory of art.

3. An explanation of the concept of distributed representation in complex systems in its application to a systems theoretical account of the institutional theory of art.

1 JoAnn Wypijewski, *Painting by Numbers – Komar and Melamid's Scientific Guide to Art* (New York: Farrar, Straus and Giroux, 1997).

1. Systems of Art: The Institutional Theory of Art from a Systems Theoretical Perspective

Systems theory provides an opportunity to rethink how art is defined. It provides an anti-ontological definition of art. In part this involves the uncontroversial step of defining art not in material terms but rather in terms of it being a function of certain social systems. In other words the systems theoretical approach suggests that rather than being an inherently differentiated set of objects and practices art gains its definition and meaning from the systems that it occurs within. Hence art is not defined by any material particularity but is rather taken to be relative to these systems of differentiation and distributed representation.

There is a cluster of philosophical discussions, based around the definition and ontological status of the work of art, which have been loosely defined as institutional theories of art. These are philosophical attempts to account for definitions of art independently from a historical teleology. It is for this reason that we can seek a common ground between such approaches and a systems theoretical approach.

Arthur Danto calls the system in which art circulates the 'artworld', which he identifies as a 'style matrix' of various influences and institutions.[2] Danto's artworld is a network of interrelated systems, which include not only the museum/gallery system but also the art market, art schools, art history, and all the protocols and conventions which form part of broader systems.

The celebrated example of Danto's argument concerns his encounter with Warhol's Brillo Boxes which he discussed first in the essay 'The Artworld' and later at length in *The Transfiguration of the Commonplace*.[3]

Whilst Danto's reflections on the institutional system of the artworld were understood as a response to Pop Art it raises a question that is transferable to the discussion on art in relation to complex sys-

2 Arthur C. Danto, 'The Artworld' (1964), in Steven Madoff (ed.), *Pop Art: A Critical Anthology* (Berkeley: University of California Press, 1997), pp. 269–278.

3 Arthur C. Danto, *The Transfiguration of the Commonplace* (Boston: Harvard University Press, 1983).

tems of distributed representation. It addresses the issue of the role context plays in the observation and representation of art; that is, in the processes by which certain objects and events are differentiated, observed and represented *as art*. Central to the argument of institutional theories of art is the claim that works of art have no physical or metaphysical differentiation from other objects or their environment other than by virtue of the way in which those objects are represented.

Danto's position can be reformulated from a systems theoretical perspective by reconsidering the role that the relevant systems of support play in the differentiation of works of art from their environment. In the case of minimalism, for example, this environment is the 'white-cube' of the gallery space. Danto argued that commonplace objects were 'transfigured' in the gallery space and, in a process analogous to alchemy, thus transformed from ordinary objects into artworks. This ontological transformation was not absolute; the representation of the objects as artworks remained relative to the system of observation.

Danto's argument was an immediate response to his personal encounter with the Brillo boxes that Warhol exhibited at The Stable Gallery, New York in 1964 (illus. 2). Danto's anxiety arose when trying to work out how these objects could possibly be seen as art when they are, at first glance no different from packages in the storerooms of a supermarket.[4] As Danto was to later recall of the moment when he encountered the boxes:

4 Nearly forty years on, Danto's shock seems anachronistic especially given that Warhol's boxes are noticeably different from their mass-produced model. Made of plywood they clearly betray the process of their crafting in the obvious material grain of the wood and the perceptible in-exactness of some of the painted design. It is hard to imagine anyone mistaking these works of craftsmanship for the hygiene and exactness of the mass-produced original. Thus Danto's observation can be seen as betraying its own historical moment in so far as Danto needs to revise his own, historically determined, idea of what art should be. Nevertheless this need not alter the efficacy of his argument which concerns the role of the gallery system in the differentiation and representation of works of art.

> It was a most exciting moment, not least of all because the entire structure of debate which had defined the New York art scene up to that point had ceased having application. *A whole new theory was called for* other than the theories of realism, abstraction and modernism which had defined the argument for Hopper and his allies and his opponents.[5]

The institutional theory of art found a further expression in Dickie's commentaries on the ontology of art[6] where he defined art as an 'institutional concept' (his terms) rather than a physical class of objects.

In Dickie's view art is thus about the way objects are classified, not about their physical qualities. Their existence is thus relative to a system of classification rather than any physical, or metaphysical, qualities. The significance lies in the way artworks are represented or performed by their system of support. By arguing in such a way Dickie rejected definitions of art in terms of expressions of either imitation or emotion. In systems theoretical terms Dickie's process of institutional 'classification' can be understood as the function of two systemic processes.

a. *Differentiation*; by which art is differentiated from its environment and

b. *Observation*; by which that differentiation is observed by the relevant systems.

Firstly Dickie argued that up to the end of the 18th century theories of art were variations of what he called the 'imitation theory of art'. He attributed such theories to Plato who had portrayed art as a deception that imitates the form of beauty in nature.[7] Likewise, Gombrich argued that art making is a process of 'making and match-

5 Arthur C. Danto, 'Pop Art and Past Futures', in *After The End of Art* (Princeton: Princeton University Press, 1997), pp. 123 f.

6 George Dickie, *Art and the Aesthetic: An Institutional Analysis* (Ithaca: Cornell University Press, 1974). Dickie's arguments have most recently been summarised in: George Dickie, *Art and Value: Themes in the Philosophy of Art* (Oxford: Blackwell 2001).

7 In Book X of *The Republic* Plato uses the example of a bed to describe three modes of reality; the ideal form, its approximation fashioned by the carpenter and finally the imitation of the imitation conceived by the artist.

ing'[8] in which artists attempt to imitate the world according to restrictions imposed by the artistic conventions they work within. In this imitation theory, which Dickie rejected, it is accepted that there is some property of the art work (in this case its ability to mimic and/or recreate a visual experience) that differentiates it from other objects.

Secondly, Dickie opposed what he called the 'expression theory of art', which he identified to have emerged with Romanticism in the nineteenth century. He argued that this conception of art called into question the empiricism and rationalism of the Enlightenment. Dickie explained Romanticism, to which the 'expression theory of art' is linked, as: 'An attempt to reach behind the sensuous screen of ordinary knowledge to something thought to be vital and important. A strong aura of religion and mysticism hovers around Romanticism.'[9]

According to the expression theory of the definition of art, art expresses (through the signs and symbols of art) the artist's feelings, emotions or intentions. Such a conception of art states that there is some property of the art work (in this case its ability to mediate and communicate an emotional experience) which differentiates it from other objects.

Dickie thus established a binary of theories of art which he rejected. On the one hand there is the materialism of the imitation theory in which art forms a special class of imitative objects. On the other hand there is the metaphysics of the expressionist theory in which art is a special class of emotive objects. In contrast to both approaches Dickie proposed an argument, which is related to Danto's arguments in both *The Artworld* and *The Transfiguration of the Common Place*, about how artworks are represented, received and defined.

Dickie acknowledged that there may be different strategies for approaching the question of defining artworks. For example a work may be identified as art by evaluative criteria that ascertain whether something is 'good enough' to be a work of art. Or, a work of art may be defined by classificatory criteria. Such judgments are related to

8 Gombrich's theory of 'Making and Matching' receives it fullest expression in: Ernst Gombrich, *Art and Illusion* (London: Phaidon, 1960).

9 Dickie, *Art and the Aesthetic*, p. 39.

how the relevant systems of classification observe and differentiate art objects from other objects based upon some of their qualities.

In summary of the above; the crux of Dickie's argument is that art is an 'Institutional Concept'. His definition of art relates it to how it is distinguished and classified relative to the system of art. Thus, the institutional theories of art present a definition of art as something that operates according to the normative function of the systems used to observe, differentiate and describe it.

Danto argued from a similar perspective when he suggested that it is through the act of classification that objects are differentiated and thus *transfigured* into art. Hence his definition of art also can be seen to be relative to the systems by which it is observed. As I outline below, those systems may behave in certain complex ways.

1.1 Semiotics, Narrative and Complex Distributed Representation

As with all other art historical and critical tendencies the institutional theory of art itself is a historically specific argument. Danto's essay 'The Artworld' first appeared in 1964 and was followed by Dickie's essay 'Defining Art' in 1969.[10] There are two points to note about this historical placing.

a. As Danto subsequently noted, his theory was a response to contemporary art movements, such as Minimalism and Pop Art, which made aesthetic capital out of questioning the environments within which they were being exhibited.[11]

b. It can be paralleled with other tendencies within the humanities to turn towards linguistics and semiotics as a mode of enquiry and structural analysis.

Both of these points draw attention to the point that at their heart the institutional theories are stoutly materialist; they argue that art

10 George Dickie, 'Defining Art', *American Philosophical Quarterly* (July 1969), pp. 253–56.

11 See Danto, *After the End of Art*, especially chapter 1, 'Introduction: Modern, Postmodern and Contemporary', in which he provided a historical account of his own arguments in relation to various 'End of Art' theories (most notably his own and Belting's).

works *are* ordinary objects and that their aesthetic and social significance is relative to the ways in which they are ordered, classified and curated. One way of reading this argument is in semiotic terms so that the concept of transfiguration is substituted for signification. The 're-signification of the commonplace' does not have quite the rhetorical punch of Danto's original but it brings the discussion back into the orbit of systems of representation. In short, the artworld might be conceived of as a representative system (in terms of a linguistic model of signification); objects within it are 'transfigured' into art by being made to signify art. Examples are urinals and (ersatz) Brillo Boxes (and suchlike) that, within the gallery system, become signs for art. Thus these objects are seen to have particular signifying functions relative to the system of representation, or narrative, within which they are observed.

However, such semiotic interpretations of the institutional theory of art are limited in so far as they operate according to the representative model of signification (mediated by the signifier/signified correlation regulated by the logic of the sign). They can be expanded by applying the systems theoretical approach to such representations. The complex model of distributed representation identifies both the contingency of the act of representation (as being relative to the system of that representation) and further identifies representation as being distributed over the entire system. In the case of the systems of the artworld this means that, instead of conceiving of particular objects as being represented as art from within particular narratives, 'art' emerges from the functioning of the system of art (and its relative subsystems).

This observation regarding the concept of *distributed* representation being used to develop the representative model of signification is a particularly important claim (and it receives a further explanation in the third section of this chapter). It surmounts a systems theoretical challenge to the structuralist basis for traditional notions of representation. In others words it is not the case that by likening institutional theories of art to linguistics or semiotics one can find a structuralist base for an analysis of the systems of the artworld. On the contrary, semiotics is not the primary model for the systems theoretical approach as I wish to apply it; and I am not seeking to subsequently embark upon a rigidly semiotic decoding of the signs and symbols of the artworld. Rather, by identifying similarities between different interpretive strategies founded upon foregrounding the significance of the representative act I want to demonstrate that language, museums, art history and other such systems are all *complex systems of distributed representation* which display isomorphic patterns of behaviour. Such complex systemic behaviour includes: that all systems become 'locked-in' around specific examples and that they are self-organising and auto-catalytic and so forth. In other words I do not claim that because both the artworld and language are systems that we can apply a linguistic analysis to the artworld, but rather that we can apply a similar systems approach to both linguistic systems and the artworld system (as well as to other complex systems).

I have argued that when the institutional theory of art is reconceived in systems theoretical terms the artworld can be thought of as a *system of art*. This means that art is represented by the complex systems of the artworld. In the chapters following this one I will explore this theme of complex representation in further detail by discussing how works of art, galleries and discourse on art can all be conceived as systems of art. However, before turning to this, I will in the remainder of this chapter focus on defining the fundamental characteristics of systems of art. In particular there are two key features that require further explanation. These are the phenomena of 'lock-in' and of distributed representation in complex systems. It is to these that I will now turn.

1.2 'Lock-In': Positive Feedback and Irreversibility

In the following section I outline the concept of 'lock-in' in complex systems in terms of positive feedback and irreversibility. 'Lock-in', is a key concept of complex systems and I use it in the chapters following this one to explain how the social systems discussed become dominated by particular cultural forms (including definitions of art and art historical methods).

W. Brian Arthur has analysed this particular aspect of the behavioural patterns of complex systems in the following terms:

> Dynamical systems of the self-reinforcing or auto catalytic type – i.e., systems with local positive feedbacks – in physics, chemical kinetics, and theoretical biology tend to possess a multiplicity of asymptotic states or possible 'emergent structures'. The initial starting state combined with early random events or fluctuations acts to push the dynamics into the domain of one of these asymptotic states and thus to select the structure that the system eventually locks into.[12]

Arthur uses the economic history of the domestic videocassette [VCR] market, and the evolutionary competition (during the period between 1975 and the mid 80s) between the two leading formats (VHS and Betamax) to show how systems become 'locked-in' to certain behavioural patterns.

The VHS/Betamax story has an interesting history. In *Fast Forward: Hollywood, the Japanese and the VCR Wars*, James Lardner briefly outlined the case study of the VCR consumer war.[13] In 1975 the Betamax format, which had been devised and was still controlled by Sony, held the monopoly on the sale of domestic VCRs. In America alone Sony sold 30,000 units during the year. However, the following year, JVC introduced the VHS format. By January of the year of the centre's completion (1977) JVC's hand was strengthened by the announcement that a further four Japanese electronics manufacturers (including RCA) had decided to construct and market VHS format VCRs. By 1978 Sony, who were still only marketing Betamax, held just 19.1% of the VCR market, compared to RCA's share of almost twice that at 36%. By 1979 VHS had become 'locked-in' as the system that lead the market and in 1981 only 25% of operating systems used the Betamax format. 'The battle [was] over' by 1987, according to an article that appeared in Rolling Stone[14] the same year. The following year, with VHS players now holding 95% of the VCR market, Sony finally conceded to manufacturing VHS formatted VCRs.[15]

12 W. Brian Arthur, 'Self-Reinforcing Mechanisms in Economics', in Phillip Anderson, Kenneth Arrow, David Pines (eds.), *The Economy as an Evolving Complex System* (Redwood City, CA: Addison-Wesley, 1988).

13 James Lardner, *Fast Forward* (New York: New American Library, 1997).

14 'Format Wars', *Rolling Stone* (Jan. 15th, 1987), p. 43.

15 'Sony Isn't Mourning the "Death" of Betamax', *Business Week* (Jan 25th, 1988), p. 37 and 'Goodbye Beta', *Time* (Jan 25th 1988), p. 52.

Using the economic system within which the sub-system of the VCR market operates, Arthur adopted the VHS/Betamax example to demonstrate an isomorphic principle which applies to the paradigms of dynamical systems theory in general; namely that through positive feedback systems may perform in a 'self-reinforcing', 'autocatalytic' and 'locked-in' manner. 'Locked-in' means that in a situation where any number of states possible for a system to achieve the system becomes phased-locked into one developmental path, regardless of the merits of the alternatives; and that this occurs through a combination of random factors, the effects of which are accentuated by positive feedback. His conclusion runs contrary to conventional economic theory, which suggests that economic systems are subject to the principle of diminishing returns. Arthur argued that some markets do in fact behave like complex systems and are subject to increasing returns. The suggestion that some observations on the complex dynamics of complex systems are creatively and meaningfully transferable (to economic systems for example) in turn opens up a Pandora's box of descriptive possibilities allowing for the discursive suggestion that other cultural systems, such as the art market, art galleries and art histories, may also be seen to behave according to complex dynamics. In other words that they also become 'locked-in' around singular foci. Arthur stated:

> Positive feedback economics, on the other hand, finds its parallels in modern non-linear physics. Ferromagnetic materials, spin glasses, solid-state lasers and other physical systems that consist of mutually re-enforcing elements show the same properties of the economic examples I have given. They 'Phase-lock' into one of many possible configurations; small perturbations at critical times influence which outcome is selected, and the chosen outcome may have a higher energy (that is, be less favourable) than other possible end states.[16]

In short, through a combination of factors the VHS format ended up gaining an upper hand in the VCR market and no-matter how slight this initial advantage was it became multiplied, through positive feedback, so that VHS ended up holding a virtual monopoly in the domestic market. This lock-in occurred because a number of factors were

16 Arthur, 'Positive Feedbacks in the Economy', in *Scientific American* (Feb. 1990).

brought into play. For example, once a customer was enticed into buying one product it was unlikely that they would buy a similar product for economic as well as pragmatic reasons, as Arthur related:

> Each format could realize increasing returns as its market share increased: large numbers of VHS recorders would encourage video outlets to stock more pre-recorded tapes in VHS format, thereby enhancing the value of owning a VHS recorder and leading more people to buy one. In this way, a small gain in market share would improve the competitive position of one system and help it to further increase its lead.[17]

There is also evidence to suggest that VHS was technically inferior and that the factors leading to its dominance may have been arbitrary – indeed Betamax is still the industry standard for video recording in filmmaking/broadcasting (although Lardner notes that JVC were the first to produce a [VHS] machine which doubled Sony's initial Betamax recording time of one hour, it may have been this advantage which swung the balance). Whatever the reasons the general principle remains; namely that in the developmental life of a complex, and non-linear system, a point of no-return, or bifurcation point, may be reached beyond which there is no turning back. The system thus becomes 'locked-in'.

2. Phase-Lock and System Attractors

The accumulative effects of positive feedback are also observable in another way. This is that small causes may produce the large effects of phase-lock, which is commonly referred to as 'butterfly effect' (e.g., the flapping of a butterfly's wings in Paris could effect the weather in New York, or small causes eventually lead to the monopoly of VHS in the VCR market). When a system has become 'locked-in' (as the VCR market had by the mid-eighties), it has settled in around an *attractor*. The apparent order can then be regarded as an emergent property of

17 Ibid.

the system. Such emergent (or macro) properties are common to the behavioural patterns of complex systems as Lewin described:

> Most complex systems exhibit what mathematicians call attractors, states to which the system eventually settles, depending on the properties of the system. Imagine floating in a rough and dangerous sea, one swirling around rocks and inlets. Whirlpools become established, depending on the topography of the seabed and the flow of the water. Eventually you will be drawn into one of these vortexes. There you will stay until some major perturbation, or change in the flow of water, pushes you out, only to be sucked into another. This, crudely, is how one might view a dynamical system with multiple attractors: such as cultural evolution, with attractors equivalent to bands, tribes, chiefdoms and states. This mythical sea would have to be arranged so that the hapless floater would be first susceptible to whirlpool one first, from which the next available would be whirlpool two, and so on. There would be no necessary progression from one to two to three to four. History is full of examples of social groups achieving a higher level of organization and then falling back.[18]

Such emergent order of the so-called *attractors* within complex systems can be visualised by means of what is called 'phase space'. Phase space maps are visual maps that can be used in the study of physical systems, such as chaotic systems, and their behaviour. On a two-dimensional graphical map the X and Y co-ordinates of a graph are used to represent the horizontal and vertical position of a point in space. But in phase space one variable of the map represents a position, while the other variable of the map represents, for example, the speed associated with the position. Gleick has given the following concise explanation of phase space:

> Any state of the system at a moment frozen in time was represented as a point in phase space; all the information about its position or velocity was contained in the coordinates of that point. As the system changed [...], the point would move to a new position in phase space. As the system changed continuously, the point would trace a trajectory. For a simple system like a pendulum, the phase space might be just a rectangle: the pendulum's angle at a given instant would determine the east-west position of a point and the pendulum's speed would determine the north-south position. For a pendulum swinging regularly back and forth, the trajectory through phase space would be a loop, around and

18 Roger Lewin, *Complexity* (New York: Macmillan, 1992), p. 20 f.

around as the system lived through the same sequence of positions over and over again.[19]

The descriptive utility of such a visual representation lies in that what is plotted is not what one would literally see since only one variable is a physical position. Instead, one is presented with a plot within the imagined phase space which depicts a general view of the behavioural patterns of the system, a view that whilst abstract, is both a suggestive and highly effective tool for studying the nature of complex systems as can be seen in the image that is taken from Gleick's book on *Chaos* (illus. 3).

As seen, the attractors are the parameters around which the system oscillates, which when visualised in a representation of Phase Space appear as the area and orbit around which the system becomes attracted. One of the reasons that this made such a large impact upon the scientific community is that, as can be seen in the fourth of Gleick's images, when plotted in phase space what at first appeared to be chaotic systems – from dripping taps to global economies – may in fact demonstrate some kind of order when visualised using phase space. This order is manifested in the characteristic butterfly shape of the fourth image, the so-called Lorenz Attractor (named after Edward Lorenz, a pioneering theorist of chaos theory).[20] This has caused many to ask the question – is there an emergent order in chaos? And if so at what stage does this order appear? From such questioning emerges the suggestion that order occurs within 'chaos' due to self-organisation within complex, adaptive systems. Order thus arises as an emergent property or a macro property of complex systems. In the chapters following this one I argue that isomorphic processes of irreversibility can be seen in the complex systems of the artworld.

As seen in the video markets and the competition between VHS and Betamx, the system reaches a point of no return beyond which it is very hard, if not impossible, for the system to return to its previous state. This point of no return, or the *bifurcation point*, is used to illustrate the irreversibility of complex system. In the video market this point of no return was the point at which VHS became the dominant

19 James Gleick, *Chaos* (New York: Vintage, 1987), p. 49.

20 Edward Lorenz, *The Essence of Chaos* (London: UCL Press, 1995).

format; and, via positive feedback, from this point gained an exponentially increasing market share over Betamax. The effect of positive feedback occurs because complex systems have memory. And because they have memory each state of the system is subsequently related to the previous system state. Hence when a system becomes 'locked-in' around an example it becomes very difficult to change the system's behaviour. I argue that when applied to the systems of the artworld that this can be used to explain patterns of cultural inertia and hegemony in systems theoretical terms.

3. Distributed Representation in Complex Systems

This third section is about distributed representation in complex systems. In short distributed representations are those which are distributed over the entire complex system and emerge from them rather than being located in particular areas of the system. In the chapters following this one I will show how this occurs in the systems of art. However to show how it operates I first look at how distributed representation functions in relation to two complex and distributed systems. I picked these two models because both have been well defined in the literature on the subject. These are:

a. *Cognitive Systems*; by which thought is modelled;

b. *Linguistic systems*; by which language is modelled.

Traditionally both cognitive and linguistic systems are conceived of as involved in the representation of the world. Systems of thought represent the world with regards to epistemological representation; and systems of language with regards to linguistic representation. Also in both cognitive and linguistic models both language and epistemology are taken to be profoundly interconnected.[21] These processes of

21 One of the most complete defences of this position can be found in Jerry Fodor, *The Language of Thought* (New York: Thomas Y. Crowell, 1975). Here Fodor advances the idea that thought is based on the symbolic logic of syntactic structures and semantic ordering. See also Chomsky's developed arguments con-

representation are, in systems theoretical terms, systems by which environments are *observed*; and likewise systems which can be observed to be involved in processes of *distributed representation*.

Further, both models provide demonstrable examples of how the application of complexity and a systems approach provides us with models which can account for the dynamic ordering and chaotic complexity of both the biological system of the human brain and the social system of linguistic expression. These examples also provide examples of where the systems theoretical perspective, or a strategy which acknowledges complexity and emergence, brings a new conceptual paradigm to proceedings: that of *distributed representation*. It is to a description of this in relation to cognitive models to which I now turn.

3.1 Two Models of Intelligence and Complex Systems: Iconographic vs. Distributed

The application of observing complexity to the problem of models of intelligence (which grew in research profile during the 1980s) provoked an ongoing and increasingly fierce debate within the Artificial Intelligence (A.I.) research community between two differing models of cognition. One is based on formal, symbolic logic as the basis for mental representation. This approach can be seen as iconographic in its outlook. The other side of the debate features a defence of cognitive models based upon complex, connected or what is called distributed representation. In short distributed representation (also called connectionism) provides a way of thinking about how the mind might work in a different way to a computer. In other words rather than representing via a system of formal symbolic logic – in an iconographic fashion, like a computer – it can be observed as a complex,

cerning the inter-relationship between thought and language; most notably in Noam Chomsky, *Syntactic Structures* (The Hague: Mouton, 1957).

dynamic and connected system which represents in a radically different manner.[22]

There are three ways to conceive of the debate regarding the impact of distribution/connectionism on cognitive modelling. These are:

a. That it can be used to replicate symbolic processing in a biologically plausible way; in this sense the 'language of thought' idea is retained but is implemented upon a distributed/connectionist system.

b. The distributed/connectionist system augments the symbolic language of thought model and fills in the gaps by providing, for example, a description of how systems might learn in a self-determining

22 Paul Cilliers provided an excellent summary of the debate on representation and distributed representation in 'Problems with Representation', in *Complexity and Postmodernity* (London: Routledge, 1998), ch. 5. He argued that the very concept of representation itself is challenged by the notion of distributed representation. However, he also retains the term 'representation' in the concept of 'distributed representation' demonstrating that whilst the concept of representation has been radically re-configured it has not been dismissed outright. See also Stephen P. Stich, David E. Rumelhart, and William Ramsey (eds.), *Philosophy and Connectionist Theory* (Hillsday, NJ: Lawrence Erlbaum Associates, 1991) and especially Tim van Gelder, 'What is the "D" in "PDP"?' and John Haugeland, 'Representational genera'. Other discussions of distributed representation in cognitive systems include: Robert M. French, 'Semi-distributed representations and catastrophic forgetting in connectionist networks', in *Connection Science* 4 (1992), pp. 365–377; Haugeland, 'The nature and plausibility of cognitivism', in *Behavioral and Brain Sciences*, 1 (1978), pp. 215–226; G. Hinton (et al.) 'Distributed Representations', in D. E. Rumelhart and James L. McClelland (eds.), *Parallel Distributed Processing* (Cambridge, MA. MIT Press, 1986), pp. 77–109; Pentti Kanerva, *Sparse Distributed Memory* (Cambridge, MA: MIT Press, 1988); D. E. Rumelhart and J. L. McClelland, 'On learning the past tenses of English verbs', in J. L. McClelland, D. E. Rumelhart, and The PDP Research Group (eds.), *Parallel Distributed Processing*, vol. 2 (Cambridge, MA.: MIT Press, 1986), pp. 216–268; Paul Smolensky, 'Connectionism, Constituency, and the Language of Thought', in B. Lower and G. Rey (eds.), *Meaning in Mind: Jerry Fodor and his Critics* (Oxford: Blackwell, 1991); S. Thorpe, 'Localized versus distributed representations', in M. A. Arbib (ed.), *Handbook of Brain Theory and Neural Networks* (MA.: MIT Press, 1995), pp. 549–552; van Gelder, 'Compositionality: A connectionist variation on a classical theme', *Cognitive Science* 14 (1990), pp. 355–384 and 'Defining "distributed representation"' in *Connection Science*, 4 (1992), pp. 175–191.

fashion but without the need to jettison the symbolic/iconographic 'language of thought' notion altogether.

c. The connectionist models replace the symbolic/logical 'language of thought' models of cognitive architecture and the concept of representation itself. This radical approach is what Cilliers argued in his attempts to provide a synthesis between connectionism and post-structuralism at the linguistic *and* epistemological levels.

Whether a final conclusion, with regards to the implications of distributed/connectionist models on cognitive modelling, can be reached is still unresolved. Nevertheless, what is clear is that the very existence of such models is forcing the issue with regards to modelling the processes of the mind in a biologically plausible manner. Such plausible models suggested by connectionism and distributed representation will, it is argued, provide accounts which accommodate the clear structural dissimilarities between the von-Neumann computing machine of G.O.F.A.I. (the symbolic processing models of so-called 'Good Old Fashioned Artificial Intelligence') in which the mind is taken to be like a computer running the software of thought, and a biological system of interconnected and distributed elements from which intelligence may be seen to emerge out of increased systemic complexity.

Regardless of the models used to achieve the ends the central aim of artificial intelligence is to provide a model of human cognition (this being the model with which we are most familiar). The aim of this is to provide both an explanation of human consciousness and the theoretical possibility of generating an 'Artificial' (that is non-human) consciousness of some description.

A fundamental question of modern, western philosophy is, as Descartes famously conceived of at the beginning of the modern age, the attempt to explain how to bridge the epistemological gulf between the internal mental life of apperception and the material world of mind independent reality. Such Cartesian dualism was ultimately epistemologically unsatisfactory and it is an uncontroversial philosophical claim to assert that the successful modelling of the mind must provide us with a physical account of mental life and that we should not rely upon positing a mind/body split with a physical body being seen as ontologically distinct from mental activity and the apperception of a metaphysical soul/sprit.

The central task of artificial intelligence is, broadly conceived, to find the physical basis of mental activity and to model it using a non-human (or artificial) technology. Thus, if we can ever make a machine 'think' then three questions are thrown up. These are:

a. What is the physical support of thought?

b. What is the nature of this thought? (Is it, for example, what Roger Penrose has called 'a very human prerogative'[23] or can it be replicated in other systems?)

c. What form does this thought take?

In addressing these three questions the act of cognition is conceived of as the act by which the world (whatever this may be) is observed by the psychic system (or the mind) of the subject.

If it feels that there has been at this point a divergence from the central theme of the chapter (art and systems) then, to clarify, my point here is to argue that this definition of complex systems, articulated via cognitive models of distributed representation (which will ultimately be applied to a systems theoretical account of art in the following chapters) finds another expression in models of how the mind works. By looking at the mind as a complex system it is proposed that an understanding of how cognitive observation and distributed representation takes place be formulated. Models of cognition that present the mind as a complex system suggest that it is a distributed system of representation and that mental activity is distributed over the whole cognitive system rather than being located in one particular area of the brain.

3.1.1 Mapping the Mind

Put very simply; the human cognitive system is a network of neurons. That's what the *soft-machine* (to use William Burroughs' term) of the brain is. These neurons are single cells which operate in conjunction

23 Roger Penrose, *The Emperor's New Mind* (Oxford: Oxford University Press, 1989), p. 3. Penrose argues that thought is peculiarly human and as a result will never be replicated artificially.

with one another by transfer of electrical current and it is by this current that the biological system of our body is able to communicate with itself and, hence, think. Each one of these neurons is connected with up to 10,000 other neurons within the system via axons which transmit information and dendrites which receive information. The meeting point between the transmitting axons and the dendrites are the synapses. And these synapses are mediated, chemically, by neurotransmitters. One neuron by itself is almost insignificant and cannot think by itself. Any significant form of mental activity requires the firing of millions of these neurons at once to generate thought.

In other words there is a physical basis for the mind and this is a complex, biological system or interconnected cells. Here are two examples which demonstrate the grounding of consciousness in the complex biological physical system of the brain.

a. The way psycho-active drugs work is by interfering with the activity of these neurotransmitters and in doing so altering thought patterns and hence perception. This provides us with an example of the impact a physical process can have on mental life.

b. Here is a more particular example.[24]

In 1998 surgeons at the University of California Medical School, using a 16 year old epileptic girl, applied a tiny electric current near to a part of her left hand side of the brain. This part was known to be related to humour. It was where Franz Gall, the founder of phrenology, had located what he called the 'Organ of Mirthfulness'.[25] The current caused the girl to laugh. When they asked her what was so funny she just said: 'you guys are just so funny – standing around.' When the current was applied again the girl found the picture of a horse in the room hilarious, and again, a third time she was amused by something else. What this example demonstrates is that a genuine human emotion, humour, can be explained by a physical cause in the complex cognitive system of the brain.

But, despite identifying a physical basis for thought as being grounded in this complex network of interconnected neurons there are

24 Quoted in Rita Carter, *Mapping The Mind* (London: Weidenfeld & Nicolson, 1998), p. 11.

25 Ibid., p.11.

various models for describing the ways in which that physical system *thinks*. As mentioned earlier representation provides us with ways of conceiving of this activity.

3.1.2 'Classical' or Iconographic Representation

When the discourse of artificial intelligence first began to emerge (in the early 1940s) the brain was taken to operate (and thus represent its environment) in the same way as a computer. This approach to cognitive modelling has been called both Classical Representation and 'Good Old Fashioned Artificial Intelligence' (G.O.F.A.I). This approach, which I will call the iconographic approach, can be critiqued from a systems theoretical perspective. The term 'iconographic representation' provides a means of distinguishing this approach with the distributed model of cognition applied in connectionist models (and also invokes a particular model of art historical reconstruction to which it will be ultimately applied).

Early attempts to 'map' the mind relied upon an understanding of computers based upon the von Neumann machine. This machine created by Hungarian mathematician John von Neumann (in the 1940s)[26] was a prototype digital computer and most computers today can be identified as conforming to the same van Neumann architecture. This architecture consists of a central processing unit which does the raw data processing of the machine according to principles of symbolic logic; that is bits of information have a symbolically representative function. This logical processing unit is the C.P.U. (central processing unit) of the computer. Secondly this C.P.U. works in parallel to an area of memory where information can be stored rather than symbolically interpreted. This area of memory can be accessed rapidly and at random and it does not need to be accessed sequentially (like information on a tape would have to be).[27] This Random Access Memory

26 For a history of the modern computer see: Paul Ceruzzi, *A History of Modern Computing: 1945–1995* (Cambridge, MA.: MIT Press, 1998).

27 Hence the difference between the von Neumann architecture and a Turing Machine which runs on a sequential tape.

(R.A.M) thus uses information in a non-semantic sense and provides it for the C.P.U. to semantically order.

The von Neumann machine works by storing instructions, or the *software*, as binary values (1s or 0s, or the states of connections being either on or off). These binary values (the 1s or 0s) are the raw data of the system and they are the instructions which are run, sequentially via the central processing unit. This data gains its symbolic, logical form relative to both its syntactic and its semantic ordering; the syntactic ordering corresponding to the order in which they're placed and the semantic ordering relating to the symbolic correspondence to their referents.

To reiterate; computers which function according to the von Neumann architecture work through the operation of a sequence of symbols presented to the central processing unit and its subsequent operation according to its particular, logical, interpretation of the symbols.[28] In effect the von Neumann machine is an iconographic system.

This G.O.F.A.I. [29] model of the sequential, logical symbolical processing machine, when applied to how the mind works leads to a conception of the mind as, in effect, a 'wet' or 'soft' von Neumann machine. A dualism is posited between hardware and software. This is the view that is opposed by the systems theoretical model of distributed cognitive representation. On the one hand there is the physical hardware which provides the architectural support for the logical operations of the system. In the computer this would be the hardware of the computer system and in the human brain the hardware is the biological soft-machine of grey matter. On the other hand there is the software or the language of thought which is the set of logical rules to be run on that hardware. This language of thought operates at a distinct logical level from the hardware. In the classical view of cognitive

28 Hillary Putnam argued for this application of cognitive modelling in *Representation and Reality* (Cambridge, MA: MIT Press, 1988), although he subsequently modified his position away from a rigidly symbolic model of mental representation.

29 In their connectionist models of cognitive architecture Rummelhart and McClelland claimed that they attempted to 'replace the "computer metaphor" as a model of the mind with the "brain metaphor"'. Rumelhart and McClelland, 'On learning the past tenses of English verbs', p. 75.

representation thought emerges because a type of software is being run on the brain.

Thus each 'thought' that the system has is taken to correspond to a particular logical function operating upon a particular part of the system. Hence each 'thought' is generated via the turning on of a bit of the system. G.O.F.A.I. relies upon there being a symbolic correspondence between thoughts and the objects of those thoughts. And in contrast to the virtual engrammes of classical representation, Smolensky[30] has observed that in the connectionist model of distributed mental representation each 'thought' generated by the system is not the result of the turning on of one bit of that system, or the implementation of a particular, discreet piece of data. Instead, in a distributed system of representation the whole system must be on and each thought is distributed over the whole of that system.

I have argued so far that cognitive models of classical representation are based upon both particular conceptions of computation and mental representation which are, essentially, iconographic in structure. The key issue here is that there is a direct correlation posited between the specific data processed by the cognitive system and mental representation. This symbolic correlation between thoughts and particular configurations of information means that classical mental representation is identified as an iconographic model of thought. Classical representation is iconographic in so far as the model is based upon the positing of a rational correlation between signifier (in this case the semantic/syntactic ordered information) and the signified (the mental representation or thought). This rational correlation is mediated by the symbol system, or the logic of the language of thought.

In this model (opposed from the systems theoretical perspective) cognitive systems are modelled as computational systems whereby thoughts are physical 'states' of the system and are expressions of a particular semantic and syntactic ordering of information. These physical states are thus the symbolic representation of particular thoughts.

30 Smolensky is one of the leading advocates of the model of distributed representation. See: Paul Smolensky, 'Connectionist A.I., Symbolic A.I. and the Brain', *AI Review*, 1 (1987), pp. 95–110 and 'On the proper treatment of Connectionism', *Behavioural and Brain Sciences*, 11 (1988), pp. 1–74.

This could be called computational in so far as Cummings observed: 'The system merits inferential characterisation because it computes representations of conclusions from representations of the corresponding premises.'[31] And, hence: 'The basic assumption of orthodox representationalism is that under proper interpretation, formal symbol crunching is cognition.'[32] In other words a thought is conceived of as a program, a set of instructions or an algorithm which is run on the hardware of the brain and is understood by virtue of the symbolic language of thought.

Another way to look at this would be to invoke the idea of engrammes and suggest that classical cognitive representation relies upon *virtual engrammes*. Engrammes are an anachronistic way of thinking about the mind as a library. (They formed the basis for Warburg's library as memory project in both his library and the Mnemosyne). Engrammes are the physical parts of the brain where particular memories would be stored. Thus if the art history section of a library burns down then we'd have no art historical memory (in that building at any rate). Similarly in the analogy of the engramme, if we lose the physical bit of the brain where we store our memory of horses then we will have no memory of horses. The classical approach is not an unsophisticated as to suggest that there is a particular physical location of every memory in the brain; it does not, for example, argue that we could point to the horse-representing part of the brain. However *virtual* engrammes form the basis of the Classical (G.O.F.A.I.) or iconographic approach to cognitive modelling. This is because specific semantic and syntactic orderings of information, or precise data-strings, correspond, iconographically, to particular thoughts.

The major problem with classical models of classical representation lies in their basis upon the engramme form. And until the distributed models of cognition based upon complex systems emerged in the 1980s models of the mind viewed the mind as a physical symbol

31 Robert Cummins, 'The Role of Representation in Connectionist Explanations of Cognitive Capacities', in: William Ramsey et al (eds.), *Philosophy and Connectionist Theory*, p. 95.

32 Ibid., p. 95.

system which was governed by rules of symbol manipulation.[33] Such models are: brittle; non-dynamic; complicated; have no common sense and they don't learn.

As a result of these shortcomings such systems are characterised by their inability to deal with language as it is spoken and understood by humans. Language is not necessarily ordered and rational but is itself a complex, dynamic and fluid system full of redundancy and ambiguity. Language is not an expert system[34] for describing the world in a discreet (i.e. non-redundant)[35] manner, and there is no direct and unchanging, iconographic, symbolic relationship between the system of language and the objects it describes. Words can mean different things at different times and in different contexts. Similarly there is not an

33 The phrase 'physical symbol system' is attributed by Rumelhart to Alan Newel. See William Ramsey et al (eds.), *Philosophy and Connectionist Theory*.

34 Expert System is a specific term which Coveney and Highfield define thus: '[a] computer program that uses a direct encoding of human knowledge to help solve complex problems, such as diagnosing an illness or interpreting the law. Also called a knowledge-based system'. Peter Coveney and Roger Highfield, *Frontiers of Complexity* (London: Faber and Faber, 1995), p. 426.

35 Redundancy has a specific meaning in terms of information-theory defined in the following terms: 'The variety in a channel that exceeds the amount of information actually transmitted. Its most common forms: (1) repetitive transmission of the same message over one channel, (2) duplication of channels, of which each could handle the transmission by itself, (3) restrictions on the use of characters or on the combinations of characters from an alphabet to form proper words or expressions (see language), (4) communicating something already known to its addressee. In the process of communication redundancy is essential to combat noise, to assure reliability and to maintain a communication channel. English writing is estimated to be 50% redundant which accounts for the ability of native speakers to detect and correct typing errors. Parity checks, which are common in communication within computers, enhance reliability but only at the expense of using additional channel capacity. The amount of information actually transmitted is not increased by this device. Similarly, Indian governments of the Mogul period are known to have used at least 3 parallel reporting channels to survey their provinces with some degree of reliability, notwithstanding the additional efforts.' From Klaus Krippendorff, *A Dictionary of Cybernetics*, an eighty page report dated Feb. 2, 1986, published online at Principia Cybernetica Web, the website of the Principia Cybernetica Project (PCP) http://pespmc1.vub.ac.be/Default.html (10/12/ 2007).

unchanging, symbolic and logical relationship between a language of thought and the content of that thought.

Classical representation is not an effective model for describing the way in which complex systems store and represent information. It is a logical, linguistic and, I would tentatively suggest, a structuralist approach to cognitive modelling.

3.2 Distributed Representation in Complex Systems

In this following (sub) section I demonstrate the concept of distributed representation in complex systems via the example of an ant colony. However distributed representation is not specific to this example and other complex systems including cognitive models and I argue the complex representative systems of the artworld also display isomorphic behaviour.

The ant colony as a complex system of distributed representation is an example which Eric Bonbeau presented at the O'Reilly Emerging Technology conference in Santa Clara in May 2002. For several years Bonabeau has been studying insect systems for examples of what he calls *Swarm Intelligence* from the Sante Fe Institute in New Mexico (which is the centre of complexity research where W. Brian Arthur amongst others' work). He's arrived at some examples of how a collective system of individually *stupid* ants (or other insects such as termites) can perform complex teleological tasks (that is tasks with a collective purpose).[36]

The example Bonabeau provided is how a (complex) system of ants may find food in a maze. Bonabeau argues that the colony as a whole will eventually find the shortest route to the food. Given a collection of hungry ants (they must, Bonabeau observes, be given a collective reason to find food) in a maze, they will all wander about in a

36 See Eric Bonabeau et al., *Swarm Intelligence: From Natural to Artificial Systems* (Oxford: Oxford University Press, 1999); and James Kennedy and Russell Eberhart, *Swarm Intelligence* (San Francisco: Morgan Kauffman, 2001), which contains (in chapter 2) a particularly useful discussion on 'swarm' systems in terms of distributed representation and applies this (in chapter 4) to observing social systems.

generally random way. This is because each single ant lacks intelligence (that is they have no individual volition to find the shortest route to and from the food and then remember that route). However, given their random, non-rational, distribution throughout the maze, then it is inevitable that some ants will find the food. Ants communicate with one another through pheromones.[37] Thus on finding the food source and wandering back through the maze they leave a trial of these pheromones for other ants to smell. Those ants to arrive at the start of the maze first are, obviously, the ones who took the shortest route and will have left the strongest scented pheromone trail as it has had less time to evaporate. The remaining ants will be naturally drawn to the strongest pheromone trail and thus the ants will be drawn back again and again to the shortest route to the food. The striking thing about this system is that it contains within it the potential to remain dynamic and can learn future routes should they present themselves. If any ants should randomly find a better route then the system can adapt to this; the way to the food isn't fixed.

The ants are a connectionist system. And despite their individual 'stupidity' they could be argued to have a collective, or 'swarm' intelligence. In *Out of Control: the New Biology of Machines, Social Systems and the Economic World*, Kevin Kelly calls this the 'hive mind'.

> Ants, too, have hive mind. A colony of ants on the move from one nest to another exhibits the Kafka-esque underside underside of emergent control… The marvel of 'hive mind' is that no one is in control, and yet an invisible hand governs, a hand that emerges from very dumb members. The marvel is that more is different. To generate a colony organism from a bug organism requires only that the bugs are multiplied so that there are many, many more of them, and that they all communicate with one another. At some stage the level of complexity reaches a point where new categories like 'colony' can emerge from simple categories of 'bug'. Colony is inherent in bugness implies this marvel. Thus there is nothing to be found in a beehive that is not submerged in a bee. And yet you

37 Pheromones are 'Chemical substances secreted by an organism that elicit a behavioural response from other organisms usually of the same species, especially substances that act as sex attractants'. E. B. Uvarov, D. R. Chapman, Alan Isaacs (eds.), *The Penguin Dictionary of Science* (London: Penguin Books, 1993).

can search a bee forever with a cyclotron and fluoroscope, and you will never find the hive.[38]

If a colony of ants is observed as a complex system then it can also be observed that the path which they trace to and from their goals are distributed representations within that system. By doing so the power of this observation becomes present. The path within the maze to and from the food is a representation of the optimum path to and from that food.

Connectionism argues that brains are isomorphic to the ant colony; that is they are both complex systems of distributed representation. A distributed system of representation represents by virtue of aspects related to its complexity; specifically these aspects are:

a. interconnection;

b. non-linearity.

Thus, isomorphic to the ant colony, a distributed system of representation is a network of interconnected elements. This interconnection is referred to as a *parallel* interconnection in so far as the separate elements can function simultaneously and do not have to work sequentially (in contrast a linear set of functions which must be executed one after each other). It is because of this simultaneous interconnectivity that the operation of the system cannot be understood as a discreet sequence of rule based activities; but rather that the way that any part of the system operates and any particular representation within the system must be conceived of in terms of the functioning of the whole of the system.

In *Frontiers of Complexity* Coveney and Highfield gave the following description of complex models of A.I.:

> Artificial neural networks mimic the brain's complexity through their non-linearity and a high degree of interconnectivity among their nodes. Like the brain, they are inherently parallel devices that can do many things concurrently. Qualitatively, their processors behave like neurons and the connections between them act like synapses. Any 'programming' of such a net consists of rules to alter the all important strengths of connections between processors. And the

38 Kevin Kelly, *Out of Control*, p. 13

> 'programs' that solve various problems emerge within such networks spontaneously.[39]

A connectionist cognitive architecture is based upon a network of interconnected nodes, or meeting points. The links between the nodes are referred to as being *weighted*. In a distributed system of representation information is shared by the whole network rather than being specifically located. Thus the information of the system is remembered and distributed over the system by the different weights of the connections between the nodes (or neurones) within the system. The *weight* of an interconnection corresponds to the amount of information that each connection carries. A weight can have a positive or a negative value depending upon whether the connection assists or inhibits the flow of information between the nodes. Thus as Cummins stated:

> The state of the network at a given time is revealed by the pattern of activation at that time, that is, by specifying the activation level of each node at that time, and by the connection weights.[40]

As a result of this distributed representation and memory; by virtue of differing weights between neurones a certain 'thought' generated by the system is observed as a pattern within the system. This pattern, or output, is a function of the weights of the connections within the network. We can think of these weights as being like the pheromone tracks of the ants with strong weights representing the shortest path to the food by a strong smell of pheromones and weak weights corresponding to a low smell.

3.2.1 Network Learning

The above example of the 'weighting' in a distributed network provides another distinction between connectionist/distributed systems and classical/iconographic systems with regards to how they 'learn'. In contrast to a connectionist system a system of iconographic representation 'learns' by being given information from an external source

39 Coveney/Highfield, *Frontiers of Complexity*, p. 302.
40 Cummins, 'Representation in Connectionism', p. 95.

in the form of a different program (algorithm) to be run on the hardware. An iconographic system treats memory as increased data in so far as there is more information provided about a particular symbolic (iconographic) representation.

To compare this with a connectionist system consider the example of the ant colony with respect to how they learnt to find their food. If, in contrast, the same problem of optimum paths to food was presented to a classical, or iconographic, system, then the representation of the optimum path would have to be presented to the system symbolically from an external source in order for the system to learn that path. It would need to be told the optimum path and then iconographically represent this path according to its own symbolic logic.

In contrast the connectionist system of the ants learns the path through self-determination. It finds the solution; or the optimum path emerges from the complexity of the system. Similarly a connectionist model of cognition (such as a neural network) treats memory in terms of the changing and irreversible structure of the whole representative system. A connectionist cognitive architecture learns not through the processing of more information but rather through an interaction with its environment during which time the environment changes the system and hence its output/patterns of distributed representation; just as more and more of the ants found the best way to get to and from the food in the maze.

In summary of the above section, as well as providing a plausible model of cognitive activity with regards the complex biological system of the brain the connectionist model also provides a plausible account of how complex systems can learn and how such learning has significant effects upon future functioning of the system. The information which is observed to be represented by a distributed network (of ants, or cognition, or the art world) at any particular time is the pattern distributed over the connections of the whole system at that particular time. Also, because distributed representation is a corollary of a pattern at a given time then that representation is observed as contingent. This means that the observed pattern that represents the system's state and hence the output of the system is dependent upon the previous state of the system. As a result of this the system can be said to have learnt according to its previous state. Hence it can be said that a complex, interconnected system of representation, by virtue of its temporal specificity and its irreversibility, can be said to have a memory.

A distributed network has memory and 'learns' by changing the weighting structure. In changing the strengths of the interactions between the nodes a different pattern will be produced given the same input by comparing and then changing the weighting pattern for any given input in relation to an expected or required output.

To reiterate; there is a clear distinction between classical and connectionist accounts of representation in relation to how these systems learn. This difference is that in:

a. Orthodox, classical, iconographic, systems: learning is just another set of rules being run *on the same system*. In other words, the architecture itself remains unaltered but the information circulating with the system changes.

b. Distributed or connectionist systems: learning, and hence producing different representational content in response to particular inputs, means that the patterns within the system are themselves altered. The ants follow a different path. The system changes by the weights of the nodes changing which means that the *system itself* changes to meet the demands of producing different representational output.[41]

3.2.2 Distributed Representation and Deconstruction

In comparing iconographic and distributed systems of representation a potential criticism of connectionism (from within the non-connectionist A.I. community) needs to be acknowledged. However, this criticism also throws its epistemological significance into relief. As Fodor,

41 Rumelhart and McClelland showed how a virtual connectionist system can be taught to learn the English past tense of verbs and convert the present into the past tense in both regular and irregular verbs. The striking claim that they made of their model was that it was given no specific rules for learning and that in the learning process it duplicated some of the mistakes which children actually make; in Rumelhart/McClelland, 'On learning the past tenses of English verbs'. This model is critiqued extensively by S. Pinker and A. Prince, 'On Language and Connectionism' and J. Fodor and Z. Pylyshyn, 'Connectionism and Cognitive Architecture', both in *Cognition*, 28 (1988).

Pylyshyn et al. argued, as a cognitive model, for all its attempts to talk of representation in terms of networks of distribution, connectionism does not account for the fact that there is a rational language of thought and that language itself, through which that thought is expressed and communicated (within social systems) is logical, symbolic and iconographic. However as Cilliers has also observed the very concept of iconographic representation is what is at stake. He wrote:

> A distributed representation is not a representation in the conventional sense of the word. It dispenses with all the components of a representational system. There are no symbols that 'stand for' something, there are no grammatical relationships between them, and the system itself has no need of a semantic level of interpretation at all. We need neither a central processor nor an outside observer.[42]

Likewise the major impact of the post-structural critique of language and rationality lies in the questioning of the direct and logical symbolic correspondence of the signifier and the signified in a representative process. Instead what is favoured is a fluid, dynamic free-play of signifiers mediated by their temporal *différence* (to use the Derridian term) from one another. Likewise distributed representation is not iconographic in its representative form in so far as there is no discreet correspondence between signifier and signified which is mediated by the structural logic of the form of the sign (as Saussure argued). Distributed representation cannot, therefore, be considered representation in the traditional sense of the word.

This connection between complexity and postmodernism is observed by Cilliers in his discussion of common-ground shared between the two discursive systems. And it is in the light of his observations regarding the isomorphisms between the systems theoretical approach and post-structuralism that my argument now turns with regards to how the art is represented in the systems of art.

42 Cilliers, *Complexity and Postmodernism*, p. 27.

4. Conclusion

I have proposed that the institutional theory of art can be reformulated from a systems theoretical perspective. This means thinking about how art might be defined in reference to the systems it is relative to. Like the institutional theory of art, a systems theoretical definition of art does not define art according to any particular qualities. Neither theory offers an ontology of art that provides a set of material characteristics to which all works of art will conform. Instead both describe how art is a product of processes of classification, differentiation and representation. However, in proposing that the artworld is a system, one can describe in more detail how the artworld operates. It operates as a complex system. In particular, like other complex systems, it displays the characteristics of 'lock-in' (positive feedback) and distributed representation. The remainder of this chapter was focused on giving models of these characteristics and explaining how they work. In the following chapters I apply these models to specific examples in the artworld. Thus a clearer picture of the systems of art emerges.

Chapter 3: Systems Aesthetics and the System as Medium

Introduction

In 1968 the artist, critic and art historian Jack Burnham made the prophetic claim that 'a *Systems Esthetic* will become the dominant approach to a maze of socio-technical conditions rooted only in the present.'[1] Burnham's claim, in a number of publications at the time, was that a burgeoning interest in systems amongst artists and writers on art would lead (in both art practice and discourse) to a paradigm shift from object to system.

In this chapter I outline the implications of Burnham's claim by exploring the key characteristics of a 'systems-esthetic' (hereafter 'systems aesthetics'). To do this I look at the ways in which a system can be conceived as an artistic medium. I do this from three main perspectives. These are:

1. The historical perspective – from which I look at how in the 1960s there was an interest in systems as both a theoretical and artistic concept and how artists began to use systems (broadly conceived) as artistic mediums.

2. The contemporary discursive perspective – from which I look at how 'system' may be employed as a key term in what Peter Osborne calls a 'retrospective critical discourse'; that is, as a means by which to rethink art practice at the end of the 1960s.

3. The contemporary practice perspective – from which I look at some examples of how contemporary practice engages in systems aesthetics through the use of system as a medium.

1 Jack Burnham, 'Systems Esthetics', *Artforum* (September, 1968); reprinted in Donna de Salvo (ed.), *Open Systems: Rethinking Art C. 1970* (London: Tate, 2005), pp. 166–169.

My central argument is that the concept of system can be used as an effective heuristic tool that provides a coherency amongst these three perspectives. It does so by demonstrating that all three positions share an investment in the idea of system as an artistic medium. Thus it allows for a wide variety of practices to be inscribed within the single rubric of system as medium. And it can therefore be applied to historical examples as well as contemporary practice and subsequently used as a means by which to map an expanded field of art displaying systems aesthetics up to the present day.

In what follows I will give a description of the historical conditions of the emergence of systems aesthetics and relate this to its reconstruction within the critical discourse of systems theory. I will conclude with some examples from contemporary practice that can be productively contextualised through the application of systems aesthetics and the idea of system as medium. In doing so, one can begin to see a pre-occupation with system as medium operating across different art historical periods and across a diverse range of artistic practices. These practices are not necessarily concerned purely with the effects of a new technology[2] (although many recent practices which explore the aesthetic possibilities suggested by new media are compatible with what I'm identifying as systems aesthetics). Systems aesthetics can also include new re-engagements with art from the end of the 1960s in which some contemporary artists have found a precedent for a socially integrated art practice that participates in social systems.

2 Edward Shanken, for example, argued that, especially in the late 20th century, 'little scholarship has explored the relationship between technology and conceptual art'. He also claimed that there was an art historical impetus to artificially distinguish information art from conceptual art. See E. Shanken, 'Art in the Information Age: Technology and Conceptual Art', in Michael Corris (ed.), *Conceptual Art: Theory, Myth and Practice* (Cambridge: Cambridge University Press, 2004).

1. The Historical Context of Systems Aesthetics

The historical interest in the aesthetics of systems between the late 1960s and the early 1970s emerged from a matrix of influences. At the time a number of key exhibitions and publications based around the theme of systems, structure, seriality, information and technology (broadly understood) took place. These are discussed in further detail below. Such exhibitions and works drew upon popular understandings of systems theory and most notably cybernetics, information theory and general system theory. They attempted to find artistic and curatorial expressions for such relatively new ideas. Significantly this interest dovetailed with three other important developments taking place at the time.

First, increased interest in the sociological and artistic significance of new technology meant that theories such as Marshall McLuhan's were increasingly explored in art and technology groups (such as E.A.T [Experiments in Art and Technology], Billy Klüver's collaboration with Robert Rauschenberg). As a result, several attempts were made to use new technology in transforming traditional object based artistic practices into new, system based ones.

Second, general social and political changes also raised interest in the notion of systems. For example the civil unrest around the year 1968 included student riots in Paris and Germany (the so-called *68er-Bewegung*); civil rights marches (e.g., in America and Ireland); assassinations in the United States (including Martin Luther King Jr on April 4, 1968); and worldwide anti-Vietnam war protests. Such civil disobedience often directed its anti-establishment and anti-authoritarian rhetoric against the 'system'. Hence systems became associated with institutional power or a prevailing social order. A similar concept of systems had also permeated the counter-culture and psychedelic movements of the 1960s. In art practice, this led to the belief that the social system could be changed by active artistic participation in it. Jack Burnham claimed, for example, that systems artists like Hans

Haacke could integrate their work into the 'real world',[3] a sentiment echoed in Haacke's own claims that his art had a political agency due to its participation in and use of systems:

> Information presented at the right time and in the right place can potentially be very powerful. It can affect the general social fabric... The working premise is to think in terms of systems: the production of systems, the interface with and the exposure of existing systems... Systems can be physical, biological, or social.[4]

Third, the pre-occupation with system as a medium had parallels in the radical art practice of the late 1960s, which both questioned and then replaced the singular art object of modernism with the 'dematerialised' art object of conceptualism, minimalism and other post-modern art practices. The outcome of these multiple influences was a number of different attempts to replace traditional media of artistic expression with the medium of systems.

1.1 Jack Burnham and Systems Aesthetics

Jack Burnham was a central, if mercurial, figure in the theorising of systems aesthetics and the idea of system as a medium at the end of the 1960s. He began his career as an artist in 1955 and until 1965 he (like Hans Haacke, his friend and collaborator) made kinetic sculptures and light based installations before he moved to art criticism. From 1968 onwards he was a contributor for *Arts* magazine and

3 'Traditionally, artworks exist in "mythical time", that is, in an ideal historical timeframe separated from the day-to-day events of the real world. Some systems and conceptual artists, such as Haacke, attempt to integrate their works in the actual events of the "real world," that is the world of politics, money making, ecology and other pursuits'. Jack Burnham, 'Steps in the Formulation of Real-Time Political Art', in Kaspar Koenig (ed.), *Hans Haacke/Framing and Being Framed: 7 Works, 1970–5* (New York: New York University Press, 1975), p. 143.

4 Hans Haacke's response (1971) to the cancelling of Haacke's show at the Guggenheim. Quoted in Lucy Lippard, *Six Years: the Dematerialization of the Art Object* 1966–1972 (Berkeley: University of California Press, 1997 [1974]), p. xiii.

became Associate Editor from 1972 to 1976. Between 1971 to 1972 he was also contributing editor for *Artforum*, which he wrote for until 1973. It is for this activity as a critic and theorist (and briefly a curator of the exhibition *Software*) that he is now most recognised. His collected essays are reprinted in *The Great Western Salt Works* (1973), which includes the two key statements of his conception of an artistic application of systems aesthetics. These are 'Systems Esthetic', which had previously appeared in *Artforum* (1968) and 'Real Time Systems' (1969).

In 1968 Burnham published the book for which he is best known, *Beyond Modern Sculpture: The Effects of Science and Technology on the Sculpture of This Century*. It is a survey of modern sculpture using vocabularies drawn from systems theory including cybernetics.[5] It concludes with a statement on the future development for art via the use of system as an artistic medium. Such sentiments are explored further in Burnham's essay of the same year in which he defined 'Systems Esthetic'. Burnham's conception of systems aesthetics was an attempt to think together, under the rubric of systems, issues regarding artistic, technological and social conditions shared by a variety of groups including artists, scientists and social theorists. It was, in part, an account of artistic responses to new technologies manifested, for example, in early computer and video art. But Burnham also noted[6] that such an artistic turn to systems thinking was a reflection of a growing interest in systems spilling over from biological and cybernetic research into open systems and communication networks found in the writings Ludwig von Bertalanffy, Norbert Weiner, Claude Shannon, Ervin Laszlo (amongst others) into society at large.

5 Burnham continued his application of systems theory as an art historical strategy in Jack Burnham assisted by Charles Harper and Judith Benjamin, *The Structure of Art* (New York: George Braziller, 1971). Shanken described it as 'one of the first systematic methods for applying structural analysis to the interpretation of individual artworks as well as to the canon of western art history itself'. E. Shanken, 'The House that Jack Built: Jack Burnham's Concept of "Software" as a Metaphor for Art', in *Leonardo Electronic Almanac*, 6 (Nov. 1998), p. 10. Also at: http://www.artexetra.com/House.html (10/08/2007).

6 J. Burnham, *Beyond Modern Sculpture – The Effects of Science and Technology on the Sculpture of this Century* (New York: George Braziller Inc., 1970), p. 12.

Such concerns with systems in the 1960s included the infamous interest in systems analysis that President John F. Kennedy and Robert McNamara brought into the United States government, largely as a strategy for modern warfare.[7] One manifestation of this was the development of Project RAND which was established by the United States Army Air Forces in 1946 and took its name from 'Research and Development'. RAND was central to the implementation of systems-analysis in military and strategic decision-making in the decades following World War II. It pioneered the development of systems analysis and its particular application for defence. As a result, in the late 1950s RAND offered short courses on systems analysis to 'senior military officers and civilians associated with the Armed Forces'.[8] These lectures were subsequently published under titles such as *Analysis for Military Decisions*[9] and advocated a systems-thinking approach to military decision making. In a publication from 1968, the use of systems thinking as a defence strategy was clearly stated as the aim of such publications:

> 'One of the key problems of contemporary national security policy', as Henry Kissinger has said, 'is the ever widening gap that has opened up between the sophistication of technical studies and the capacity of an already overworked leadership to absorb their intricacy'. This book, a survey of the nature, aims and limitations of systems analysis in current defense planning, is an attempt to close that gap. We focus on systems analysis because it is unquestionably the most powerful and widely influential approach to systematic inquiry that decision makers and policy-orientated analysts have at their disposal today – and are likely to have in the foreseeable future.[10]

7 In 'Systems Esthetics' Burnham makes reference to the use of systems theory by the Pentagon. For a fuller discussion on this topic see Paul Dickson, *Think Tanks* (New York: Ballantine Books, 1971).

8 See 'Preface' to E.S. Quade and W.I. Boucher (eds.), *Systems Analysis and Policy Planning: Applications in Defense. A report prepared for United States Air Force, Project Rand* (New York: American Elsevier Publishing Company, June 1968), p.v. At: http://www.rand.org/pubs/reports/R0439/ (10/08/2007).

9 E.S. Quade (ed.), *Analysis for Military Decisions* (Chicago: Rand McNally & Company, 1964).

10 Quade/Boucher, *Systems Analysis and Policy Planning*, p. v.

And,

> Since 1961, the United States has introduced a new philosophy, technique and style of defense management… In its research for the United States Air Force, The RAND Corporation has played a leading role in developing an approach to the full range of these problems and in bringing the methods used to national attention. This approach, which we call 'systems analysis', is the subject of this book.[11]

The potential for systems analysis was enthusiastically pursued as a new paradigm in social thinking. For example, Norbert Weiner's development of cybernetics had an early manifestation in his collaboration with Julian H. Bigelow on anti-aircraft defence systems in World War II (although he was later to move to more humanist applications of cybernetics); and in 1967 M. Ways could write the following optimistic analysis of the efficacy of systems thinking as a military and strategic decision-making tool to implement 'world system' policy decisions which reads chillingly today:

> That the US still retains its limited-war option is due in no small measure to the confidence that a large part of the public and the white-house have in McNamara's system of planning; it is deemed sensitive enough and effective enough to maintain a level of operations in Vietnam calibrated to our limited objectives there, and to changing circumstances. True there have been shortages of specific materials and underestimations of cost. Nevertheless, the Vietnam war thus far has been the best calculated military supply effort in 20th century US History. If McNamara's planning system successfully demonstrates in Vietnam that the US can maintain an adequate, though limited, response then the consequences for foreign policy in the next ten years may be profound.[12]

Burnham was familiar with such applications of systems thinking. In 'Systems Esthetics' he cited the following directly from E.S. Quade's essay, 'Methods and Procedures' in *Analysis for Military Decisions*: 'Systems Analysis, particularly the type required for military

11 Ibid., p. 1.
12 M. Ways, 'The Road to 1977', *Fortune* (Jan., 1967), pp. 93–95; 194–197 (Time Inc.).

decisions, is still largely a form of art. Art can be taught in part, but not by the means of fixed rules.'[13]

The use of the citation by Burnham presents a problematic conflation of military strategy with art making. It also, however, demonstrates Burnham's equation of systems thinking with a certain type of aesthetic sensibility as will be discussed in further detail below.

1.2 Exhibitions & Publications

By the end of the 1960s the interest in the application of systems thinking by the military-industrial complex began to filter into cultural life. Between the years 1966 and 1972 there were a number of important exhibitions and publications that took the idea of systematicity as their central organising principle, with titles such as *Systems; Information*; *Software* and *Radical Software*.

For example, Kynaston McShine's (with Lucy Lippard) *Primary Structures* show in 1966 (The Jewish Museum, NYC) was a key exhibition of forty artists (including Carl Andre, Walter de Maria, Dan Flavin, Robert Smithson and Sol Lewitt) that documented the crystallisation of minimalism through the investigation of simplified sculptural form. Importantly, however, the exhibition emphasised the systematic structures that underlay such minimal forms. In a similar way, Harald Szeemann's *When Attitudes Become Form: Concepts, Processes, Situations, Information* (Kunsthalle Bern/ Institute of Contemporary Art, London, 1969) explored sculptural possibilities of post-minimalism, while its subtitle invoked systems thinking, albeit implicitly, by using vocabularies grounded in information theory and communication theory. In 1969 the British artist Jeffrey Steele organised an exhibition of English artists at the Amos Anderson Museum, Helsinki, under the title *Systems*, focusing on abstract art. Early the following year Steele and another artist, Malcolm Hughes, formed a group that focused on

13 'Methods and Procedures', from Quade, *Analysis for Military Decisions*. For a further discussion on the use of computers in warfare in the 20th Century see P. Edwards, *The Closed World: Computers and the Politics of Discourse in the Cold War* (Cambridge, MA: MIT Press, 1996).

the discussion, promotion and exhibition of systems art. The main outcome of this group activity[14] was the exhibition *Systems*, organised in conjunction with the Arts Council, which opened March 1972 at the Whitechapel Gallery in London. The exhibition focused upon how both system and structure (mostly mathematically understood) could be used as a foundation for abstract art.

In addition, there were also a number of exhibitions that took the intersection of art and technology as their central theme.[15] These included *Cybernetic Serendipity: The Computer and the Arts* (I.C.A., London, 1968) curated by Jasia Reichardt.[16] Reichardt deliberately invoked Norbert Wiener's definition of cybernetics as a title around which to co-ordinate works that, as was claimed in the press release:

> ... are either produced with a cybernetic device (computer) or are cybernetic devices in themselves. They react to something in the environment, either human or machine, and in response produce either sound, light or movement.[17]

The Machine as Seen at the End of the Mechanical Age (Museum of Modern Art, New York, 1968) curated by K.G. Pontus Hultén also explored the connections between art and technology. It included works from over a hundred artists drawn from the fifteenth to the twentieth century and presented ways in which technological systems historically featured as part of artistic media. Amongst the works shown were drawings by Leonardo Da Vinci of flying machines, the first artistic exhibition of a video cassette recorder used by Nam June Paik and the winning entries to a competition organised by E.A.T.

14 The artists involved were: Richard Allen, John Ernest, Malcolm Hughes, Colin Jones, Michael Kidner, Peter Lowe, James Moyes, David Saunders, Geoffrey Smedley, Jean Spencer, Jeffrey Steele and Gillian Wise Ciobotaru.

15 For a fuller discussion of the various manifestations of the art and technology movement see Marga Bijvoet, *Art as Inquiry: Toward New Collaborations Between Art & Science* (Oxford: Peter Lang, 1997).

16 For a full historical discussion of the show see: Rainer Usselmann, 'The Dilemma of Media Art: Cybernetic Serendipity at the ICA London', *Leonardo*, vol. 36, no. 5 (Oct. 2003), pp. 389–396.

17 The press release for the exhibit curated by Reichardt is quoted from: *Media Art Net*: http://www.medienkunstnetz.de/exhibitions/serendipity/#reiter (10/10/07).

(Experiments in Art and Technology) promoting the collaborations between artists and engineers using new technology.[18]

Such themes were mirrored in Kynaston McShine's international show *Information* (Museum of Modern Art, NYC, Summer, 1970), which Lucy Lippard described as follows:

> Born of an art-orientated interest in systems and information theory, and then transformed by the national rage attending Kent State and Cambodia, it became a state-of-the-Art exhibition unlike anything else that cautious and usually unadventurous institution had attempted date. The handsome catalogue looked like a conceptual artists book, with its informal 'typewritten' text and wild range of non-art imagery from anthropology to computer science, and an eclectic, interdisciplinary reading list.[19]

Hans Haacke's 'MOMA Poll' (1970) was included in the exhibition. It was one of the several 'Visitor's Profiles' projects Haacke conducted between 1969 and 1970 to collect statistical information from various galleries. For the installation at *Information* the 25,566 participants who took part voted either 'yes' or 'no' to the question, 'Would the fact that Governor Rockerfeller has not renounced President Nixon's Indochina policy be a reason for you to not vote for him in November?' The results were subsequently collated by an electronic counting device and displayed.

The aforementioned exhibitions were important public expressions of the emerging systems aesthetics. They were not, however, the only expressions. The journal *Radical Software*, for example, also explored the intersections between technological systems and art. Eleven editions were published by The Raindance Corporation between 1970 and 1974. The Raindance Corporation was formed in New York in 1969 by the artist Frank Gillette. As Davidson Gigliotti recounted:

18 E.A.T. followed from the event *Nine Evenings: Theatre and Engineering*, organised by Robert Rauschenberg and Billy Klüver at the Armoury Building, New York City, 13–22 October, 1966 to promote the collaboration between artists and engineers. They also organised the Pepsi Pavillion at the World's Fair, Osaka, in 1970. For a detailed discussion of the project see Bijvoet, *Art as Inquiry*, ch. 2.

19 Lippard., *Six Years*, p. xix.

> It was Gillette's intention to found an alternative media think tank; a source of ideas, publications, videotapes and energy providing a theoretical basis for implementing communication tools in the project of social change. To make his point, Gillette chose the name Raindance as an ironic reference to the Rand Corporation, then and now an establishment think tank advising government and industry.[20]

Staff at *Radical Software* included Michael Shamberg (publisher), Ira Schneider (co-originator) as well as Beryl Korot and Phyllis Gershuny (editors). *Radical Software* was concerned primarily with the artistic possibilities of low-cost portable video equipment (such as the Sony CV Portapack,[21] which had become available in 1965). Contributions included those by Nam June Paik, Douglas Davis, Paul Ryan, Frank Gillette, Beryl Korot, Charles Bensinger, Ira Schneider, Ann Tyng, R. Buckminster Fuller, Gregory Bateson, Gene Youngblood, Parry Teasdale and Ant Farm. Although also concerned predominantly with the artistic and social implications of new technology, the journal also frequently included polemic articles that fall within the orbit of a 'systems' outlook. As Bijvoet noted, Michael Shamberg actively engaged in the use of 'new lexicons' such as:

> Cybernetics and systems notions with their accompanying vocabularies were mainly applied to the possibilities of new media systems, such as video, cable,

20 Davidson Gigliotti, 'A Brief History of RainDance' (2003) on the website for *Radical Software* http://www.radicalsoftware.org/e/index.html (10/09/2007).

21 'In 1963, the very first home videotape recorder appeared in the Nieman-Marcus Christmas catalog. It was from Ampex; it was called the Signature V; it cost $30,000 (...) It was the size of a coffin; it weighed more. (...) Sony, active in the industrial video arena for years, introduced its CV series half-inch, black/white open-reel format in 1965. (...) 'CV' ostensibly stood for 'consumer video', and machines actually were sold to home users in such big-ticket emporiums as Neiman-Marcus. The first CV machine (which weighed in at a mere 70 pounds) even had a built in nine-inch monitor that popped up for viewing. The format initially produced jittery, flickering images, but incorporated some features that later became well loved, such as timer recording. Although it didn't make much of a splash in the stores, CV made it into some school systems. One [Video Review] editor remembers making his television debut on his grammar school's closed circuit TV channel, which employed CV equipment. By the end of the 60s, Sony went back to the drawing board.' From *Video Review* (April 1991), pp. 32, 34–35.

satellite, etc. Words like system, feedback, information, software, parameter, entropy and negentropy, process, pattern became the principal vocabulary in the writers' argumentations.[22]

These include, for example, an article by Paul Ryan declaring the contemporary need for an 'information economy' in issue 3 (Spring, 1971), and pieces by Gregory Bateson and Gene Youngblood on ecologies of media and media systems.

1.3 Software and Systems Aesthetics

In the midst of this curatorial and publishing activity was Burnham's own exhibition *Software* (The Jewish Museum, 1970) which to date has been the only curatorial expression of his notion of systems aesthetics. As Edward A. Shanken noted, the uniqueness of the show lay in its attempt to express thoroughly Burnham's concept of systems aesthetics:

> In contrast to the numerous art and technology exhibitions which took place between 1966–1972, and which focused on the aesthetic applications of technological apparatus, *Software* was predicated on the ideas of 'Software' and 'Information Technology' as metaphors for art. He conceived of 'software' as parallel to the aesthetic principles, concepts or programs that underlie the formal embodiment of the actual art objects, which in turn parallel 'hardware'.[23]

The show contained a number of potentially innovative exhibits exploring ideas of systematicity, interactivity; the use of new technological systems in art making and the shift from singular art objects to systems of art. Several of the works were intended to be controlled by a DEC PDP-8 Time Share Computer (which, Shanken noted, failed to work for the first month.) Many works were conceived of as completely interactive with full visitor participation.

Paradigmatic works from the show included 'Labyrinth' by Ted Nelson and Ned Woodman, described by Shanken as a 'hypertext system' in which:

22 Bijvoet, *Art as Inquiry*, p. 75.
23 See Shanken, 'The House that Jack Built'.

> Users could obtain information from a 'interactive catalogue' of the exhibition by choosing their own narrative paths through and interlinked database of texts, then receive a print-out of their particular 'user history'.[24]

Les Levine's 'Systems Burn-Off X Residual Software' was also intended to be fully interactive:

> The original installation at the Phyllis Kind Gallery in Chicago was comprised of 1,000 copies of each of 31 photographs taken by Levine at the March, 1969 opening of the highly publicized 'Earthworks' exhibition in Ithaca, New York ... Most of the 31,000 photographs, which documented the media event, were [according to the catalogue] 'randomly distributed on the floor and covered with jello; some were stuck to the wall with chewing gum; the rest were for sale'... Levine conceived of the 31,000 individual photos as the residual effects or 'burn-off' of the information system he created – as the material manifestation of software.[25]

There were also two pieces contributed by Hans Haacke: 'News' (illus. 4) and 'Norbert: "All Systems Go."' 'News' presented several Teletype machines upon which reports of local, national and international news spilled out on continuous rolls of paper into the gallery. It thus brought communications systems directly into the gallery space and encouraged direct visitor participation. 'Norbert: "All Systems Go"' (illus. 5) on the other hand was another of Haacke's engagement with 'Real-Life Systems' that is work produced with live animals or biological organisms. In this case a mynah bird, sardonically named after Norbert Weiner, was trained to say 'all systems go'.

As Burnham explained, *Software* was:

> ...an attempt to produce aesthetic sensations without the intervening 'object'; in fact, to exacerbate the conflict or sense of aesthetic tension by placing works in mundane, non-art formats.[26]

As the rationale and exhibits in the show demonstrate he was able to do this by invoking the concept of 'software' as both a central

24 Ibid.

25 See Shanken, 'Art in the Information Age'.

26 Burnham in personal correspondence with Edward Shanken quoted in Shanken, 'The House that Jack Built', p. 2.

dynamic and metaphor for an interactive art practice. To thus understand art as software is to invoke the paradigm of the coding of a computer program which is not copied in different hardware, but is rather given other manifestations. Les Levine, for example, saw his contribution to *Software* in these specific terms: and wrote the following description of 'Systems Burn-off X Residual Software' in the *Software* catalogue: 'In many cases an object is of much less value than the software concerning object. The object is the end of a system. *The software is an open continuing system.*'[27]

More recently, Lev Manovich argued that the application of the metaphor of 'software' can be extended beyond computers to a definition of art since the advent of modernism in general:

> To summarize: from the new vision, new typography, and new architecture of the 1920s we move to new media of the 1990s; from 'a man with a movie camera' to a user with a search engine, image analysis and visualization programs; from cinema, the technology of seeing, to a computer, the technology of memory; from defamiliarization to information design. *In short, the avant-garde becomes software.* This statement should be understood in two ways. On the one hand, software codifies and naturalizes the techniques of the old avant-garde. On the other hand, software's new techniques of working with media represent the new avant-garde of the meta-media society.[28]

Manovich, like Burnham, was looking for an adequate vocabulary to describe a variety of diverse artistic practices. Both recognised the potential in systems theory to reframe a discussion of the avant-garde. For Burnham in particular this entailed a formulation of an aesthetic theory based upon the central paradigm of the possibility for a system to be conceived as a medium for art. A further discussion of what this means from a contemporary perspective follows below.

27 Quoted from N. Wardrip-Fruin and N. Monftort, *The New Media Reader* (Cambridge, MA: MIT Press, 2003), p. 255; emphasis added.

28 Lev Manovich, 'Avant-garde as Software', *Artnodes*, http://www.uoc.edu/artnodes/eng/art/manovich1002/manovich1002.html (22/03/2006); emphasis added.

2. Systems Aesthetics as a Retrospective Critical Discourse

To talk about art in terms outlined by Manovich and Burnham, that is, as software and system, is to invoke terms from cybernetics as well as information and communication theory. It means, therefore, to talk in terms developed in the 1960s and thus exclude the later configurations of systems theory (such as the social systems theory developed by Luhmann) that I wish to employ in my account.

To understand art as software is to understand it in terms of codes and information rather than in material or medium-specific terms. Burnham proclaimed that systems aesthetics necessitated the dissolution of the material specificity of traditional artistic mediums so that the traditional 'objet d'art' would eventually be replaced by 'aesthetic systems'.[29]

> [The] cultural obsession with the art object is slowly disappearing and being replaced by what might be called 'systems consciousness'. Actually, this shifts from the direct shaping of matter to a concern for organizing quantities of energy and information.[30]

In his advocacy of 'the organizing quantities of energy and information' for a new type of art making Burnham advanced an art that was both ontologically unstable and interactive. He thus specifically linked a post-formalist artistic attitude prevalent in the art world at the time that was also interested in exploring de-ontologized and interactive art to the contemporary discourse of systems theory. In 1968 he wrote:

> The post-formalist sensibility naturally responds to stimuli both within and outside the proposed art format… [but] the term systems esthetic seems to encompass the present situation more fully.[31]

29 'The substitution of "aesthetic systems" for the objet d'art within the confines of a gallery is something that should be fully developed in another book.' Burnham, *Beyond Modern Sculpture*, p. 11.

30 Ibid., p. 369.

31 Burnham, 'Systems Esthetics', *Artforum* (Sept. 1968).

In doing so Burnham articulated in systems theoretical terms the emerging historical situation in art practice toward the end of the 1960s. We need not, however, be constricted by Burnham's definition alone. Systems aesthetics can also operate as what Peter Osborne called a 'retrospective critical discourse', which 'does not need to discover its terms literally or empirically within the discourse of the period under discussion'.[32] This means that systems aesthetics can also be identified as a function of the discursive system from which it is observed and constituted; it thus can be integrated into a coherent historical and sociological narrative.[33] It is from this perspective that I wish to continue.

It is my argument here that Burnham's systems aesthetics is compatible with a variety of art historical descriptions and can therefore be employed as part of a retrospective critical discourse of systems aesthetics. Thus conceived, systems aesthetics allows for an expanded field of practice implying a shift from singular art objects to the use of systems as artistic mediums. These descriptions include: the 'dematerialised' art object (Lucy Lippard); 'Intermedia' (Dick Higgins); and the 'post-medium condition' (Rosalind Krauss).

Lucy Lippard's definition of the 'dematerialised' art object highlighted what was at stake when the anti-modernist aesthetic lead to a dismantling of the modernist art object. In 1973, looking back on the immediate past, she observed that the six years following 1966 had

32 Peter Osborne, 'Presentation', *Open Systems: Rethinking Art c. 1970 Symposium* (16–17 Sept., 2005). Archived at http://www.tate.org.uk/online events/archive/OpenSystems/#friday (11/10/2007). Osborne continued by saying that any attempt to 'literally or empirically' reconstruct the terms of one's discourse from the historical period is a 'phantasmatic illusion' because the 'criteria of validity for critical discourse are different from those of empiricist historiography'.

33 And acknowledging this means recognising the relationship between the discursive position of systems theory and the historical phenomenon that it both observes *and constitutes* by virtue of that observation. Luhmann engaged specifically with the issue of discursive self-reflexivity and the contingent relationship between an observing discursive system and that which it observes. See Niklas Luhmann, 'Deconstruction as Second-Order Observing', *New Literary History* 24, pp. 763–782. See also William Rasch, *Niklas Luhmann's Modernity* (Stanford: Stanford University Press, 2000).

been characterised by what she termed, 'the dematerialisation of the art object'. Lippard related how an interest in singular objects, adhering in specific mediums, was replaced by work which explored its relationship with its various systemic environments. Her micro-history of the six years at the end of the 1960s focused, in her terms:

> ...on so-called conceptual or information or idea art with mentions of such vaguely designated areas as minimal, anti-form, systems, earth or process art, occurring now in the Americas, Europe, England, Australia, and Asia (with occasional political overtones).[34]

At the heart of these activities lay a series of practices which radically challenged the faith in an ontologically stable, modernist art object; one that subsisted in a specific medium. In doing so it operated according to a self-aware artistic practice that placed the questioning of the relationship between a work of art and its various environments at the very centre of the work's meaning. Such work can be understood as exploring an aesthetics of systems and in doing so thus functioned by investigating the ways in which it was embedded in various networks of display, representation, meaning and control.

Dick Higgins used the term 'intermedia' in 1966 to explain the same historical situation that he described as: 'much of the best work being produced today seems to fall between media. This is no accident.'[35] He observed that the use of a wide variety of media in, for example: conceptual art; mail art; performance art (and so forth) was a means by which art of the age (which he called the 'third industrial revolution') would identify its distinction from that of the art of the Renaissance. Given its plurality, intermedia art is resistant to an account of it in terms of the material specificity required of it by a limited account of modernism. Thus, the term 'intermedia art', is, like Lippard's phrase, compatible with Burnham's account of a practice

34 See Lippard, *Six Years*.

35 Dick Higgins, 'Intermedia' (1966), reprinted in Donna De Salvo (ed.), *Open Systems Rethinking Art c. 1970* (London: Tate Publishing, 2005). See also Dick Higgins, *Horizons, the Poetics and Theory of the Intermedia* (Chicago: Illinois University Press, 1984) and Y. Spielmann, 'Intermedia in Electronic Images', *Leonardo*, 34.1 (2001), pp. 55–61.

that attempted to make art without the production of unique art objects.

The condition of the dematerialised or intermedia art after modernism was called the post-medium condition by Rosalind Krauss. Krauss grounded her discussion in the specific historical role that the technological development of photography played in post-war artistic practice. In particular she focused on the convergence of art and photography in the 1960s. Photography, she argued with Benjamin, brought about a challenge to the unique status of art by challenging the status of the specific aesthetic object of art. Krauss argued that this challenge plays out in a lack of faith concerning medium specificity and in a 're-invention' of the conceptual coupling between modes of artistic production and the media within which those modes are enacted. It is worth quoting Krauss at length on this issue:

> This time, however, photography functions against the grain of its earlier destruction of the medium, becoming, under precisely the guise of its own obsolescence, a means of what has to be called an act of reinventing the medium. The medium in question here is not any of the traditional media – painting, sculpture, drawing, architecture – that include photography. So the reinvention in question does not imply the restoration of any of those earlier forms of support that the 'age of mechanical reproduction' had rendered thoroughly dysfunctional through their own assimilation to the commodity form. Rather, it concerns the idea of a medium as such, a medium as a set of conventions derived from (but not identical with) the material conditions of a given technical support, conventions out of which to develop a form of expressiveness that can be both projective and mnemonic. And if photography has a role to play at this juncture, which is to say at this moment of postconceptual, 'postmedium' production, Benjamin may have already signalled to us that this is due to its very passage from mass use to obsolescence.[36]

Krauss thus argued that post-medium art required a jettisoning of a certain modernist understanding of medium specificity most commonly associated with Clement Greenberg:

> At first I thought I could simply draw a line under the word medium, bury it like so much critical toxic waste, and walk away from it into a world of lexical

36 Rosalind Krauss, 'Reinventing the Medium (art and photography)', *Critical Inquiry*, vol. 25, no. 2 (Winter, 1999), p. 289.

freedom. 'Medium' seemed too contaminated, too ideologically, too dogmatically, too discursively loaded... That such a definition of the medium as a mere physical object, in all its reductiveness and drive toward reification, had become common currency in the artworld [via Greenberg]. Indeed so pervasive was this drive to Greenberg-ize the word that historically previous approaches to its definition were now stripped of their own complexity.[37]

Krauss, however, did not wish to completely dismiss the concept of medium and advocated instead a definition of medium as the 'technical support' for the work of art. The benefit of this expanded definition of medium is that it does not make medium reducible to, 'the specific material support for a traditional aesthetic genre'. As Krauss argued, such an expanded definition of 'technical support' is beneficial:

'Technical support' has the virtue of acknowledging the recent obsolescence of most traditional aesthetic mediums (such as oil on canvas, fresco, and many sculptural materials, including cast bronze or welded metal), while it also welcomes the layered mechanisms of new technologies that make a simple, unitary identification of the work's physical support impossible (is the 'support' of film the celluloid strip, the screen, the splices of the edited footage, the projector's beam of light, the circular reels?) If the traditional medium is supported by a physical substance (and practiced by a specialized guild), the term 'technical support', in distinction, refers to contemporary commercial vehicles, such as cars or television, which contemporary artists exploit, in recognition of the contemporary obsolescence of the traditional mediums, as well as acknowledging their obligation to wrest from that support a new set of aesthetic conventions to which their works can then reflexively gesture, should they want to join those works to the canon of modernism.[38]

The idea that medium be understood as the 'technical support' for a work of art means that its definition is grounded upon a set of historically situated practices rather than a particular set of material conditions. Thus painting, for example, is not explicable by the existence of paint, canvas and brushes alone, but instead must be understood as the set of historical conditions that allow for the identification of a set of technical procedures (such as historical precedents and

37 Ibid., p. 5.

38 Krauss, *A Voyage on the North Sea: Art in the Age of the Post-Medium Condition* (New York: Thames & Hudson, 2000), pp. 55–62.

studio practices) to be acknowledged as painting.[39] This is, crucially, a description of medium that is expansive enough to include system as a medium.

Krauss' notion of the post-medium condition takes account of a situation in the 1960s, which was also described by Lippard, Higgins and, crucially Burnham. This was a situation when artists began to critique ideas of modernist medium-specificity. The systems aesthetic thus provides the opportunity to re-inscribe materially diverse practices (such as minimalism, conceptualism and new media art) within the single medium of system. In doing so the 'retrospective critical discourse' of systems theory provides the vocabulary to map different artistic strategies within a single rubric. One may thus discuss historical examples as contiguous with more recent ones. And it is to these that I turn in the concluding section.

3. Systems Aesthetics and Contemporary Practice

In this brief concluding section I reflect upon the implications of the paradigm of the systems aesthetic to reflect upon contemporary practice. It has been my central claim in this chapter that the systems aesthetic employs the idea that system be understood as an artistic medium. I hope to have shown that when system is understood as a medium for art that this provides a medium that is flexible enough to describe a materially and historically diverse set of practices.

39 The example of painting as a medium in Krauss' terms is illustrated in Courbet *Painter's Studio* (1854–55) which Courbet himself identified as a 'real allegory' of the technical and historical conditions of the medium of painting at the time at which it was painted. Michael Fried has developed this further by arguing that the painting is (with *The Wheat Sifters*, 1853) an 'allegory of its own production' in which Courbet depicts himself as 'already immersed' in his medium and 'physically enclosed, one might say subsumed, within the painting he is making, wherever the ultimate limits of that painting are taken to lie'. Michael Fried, *Courbet's Realism* (Chicago: University of Chicago Press, 1990), pp. 155–161.

Recent curatorial interest in systems has demonstrated both its effectiveness as a descriptive paradigm and its relevance to contemporary practice. For example in the exhibition *Open Systems: Rethinking Art c. 1970*, held in 2005 at Tate Modern, the curator Donna de Salvo used the concept of system as an organising principle to arrange art from a few years either side of 1970. The show demonstrated that as an organising principle system was flexible and suggestive enough for 'rethinking' historical practice whilst also providing a focus that is effective, rigorous and engaging (and popular.) The concept of system thus provided an opportunity to re-evaluate the chosen work in a broader historical and aesthetic context in which the often unhelpfully narrow titles of minimalism, conceptualism and so forth could be sidestepped in favour of critical descriptions grounded upon a shared interest in systems of display, representation, meaning and control. The art historical benefit of this was demonstrated by the exhibition uncovering interesting connections between artists who produced work as visually diverse as Robert Smithson, Carl Andre, Lygia Clark, Andy Warhol and Bruce Nauman. To reflect this systemic connection De Salvo explained in the accompanying catalogue that all works on display were, 'linked by their use of a generative or repetitive system as a way of redefining the work of art, the self and the nature of representation'.[40] She further explained that the most interesting work from the late 1960s and early 1970s, both historically and in terms of its influence on contemporary practice, are minimalism, conceptual art fluxus and neo-concretism. Da Salvo's claim on the persisting relevance of systems art is supported by three instances of contemporary curatorial and artistic interest in systems.

First, there was the appearance of several shows between 2004 and 2006 that concentrated on a reappraisal of art from the late 1960s and early 1970s. These included several high profile retrospectives of Donald Judd (London, Düsseldorf, Basel) Dan Flavin (Washington, Forth Worth, Chicago) Robert Smithson (in Los Angles, Dallas and

40 Donna De Salvo, 'Where We Begin – Opening the System, c. 1970', in Donna De Salvo (ed.), *Open Systems Rethinking Art.*

New York) all of which were accompanied by both critically important catalogues and several definitive publications such as *Donald Judd: Complete Writings* (Nova Scotia Museums, 2005) and *Robert Smithson: Spiral Jetty* (University of California Press, 2005). This was followed by contemporary installations by artists who had been active since the 1960s, including Bruce Nauman's *Raw Materials* (2004–2005; Tate Modern) which served as a type of audio retrospective of his work and Richard Serra's 2006/07 installations at the Guggenheim in Bilbao and New York. All this activity re-enforced the cues that contemporary artists (for example Liam Gillick, Carsten Höller and Mark Dion) were clearly taking from conceptualism, minimalism and institutional critique; and thus demonstrated how systems could be employed to map an expanded field of practice from historically different times.

Second, such a mapping would draw direct connections with other artists who have specifically investigated how their own bodies are situated within various social, physical and psychic systems. An exemplary artist is Stellarc. In his bodily modifications he subordinates his physical body to cybernetic and technological systems. For example in *Ping-Body* performance (10 April, 1996, 8.00 pm at Artspace, Sydney, Australia, as part of the Digital Aesthetics Conference) his body was controlled by prosthetic extensions remotely controlled by users over an internet connection. Another, very different, example is Brian Eno's *77 Million Paintings* (Laforet Museum, Harajuku, Tokyo, Japan, 2006) for which he used self-generating and developmental algorithmic systems to create a sequence of evolving patterns or 'paintings' displayed on monitors and television screens. These examples, although this is of course not an exhaustive list, demonstrate the diversity of art practices whose predominant mediums can be observed as systems.

Third, renewed contemporary interest in self-critical practice that expands its scope beyond the limits of a singular art work into the social systems within which it is placed has been the specific tenet of French curator Nicolas Bourriaud's sometimes controversial practice and of his concept of 'relational aesthetics'. Bourriaud has attempted to engage with artistic practice since the 1990s (again, the examples of Liam Gillick, Carsten Höller and Mark Dion are appropriate) and has pushed for a move into 'the realm of human interactions and its social context, rather than the assertion of an independent and private symbolic space'.[41] Such work arguably takes its cues from the rigorous probing of the status of art and its institutions that is characteristic of art that engages in systems aesthetics (for example Burnham and Haacke's work from the late 1960s and early 1970s). In his *Relational Aesthetics*, Bourriaud argued that art institutions might become more than mere conditions of production and display. They are observable, he maintained, as part of the social systems that create new venues for relations among different practices and viewers; they are both 'frame' for the work, and part of its medium. Relational aesthetics is thus compatible with systems aesthetics in so far as it radically reconceives the purposes and effects of art practice and thus puts into question common notions of the nature of art objects. This reconceived understanding locates art in a system of relationships between art and: its environment; its viewers and art discourse.[42]

3.2 New Media, Systems and Medium

I close with a further example of how the notion of system might be usefully employed by contemporary art historians. This is in providing an account of so-called 'new media art'. New media art has had a problematic reception in art historical terms. As Charlie Gere observed, it raises the question as to what critical discourse is supposed to deal with it: 'If new media art wishes to be taken seriously then it is necessary to start to develop appropriately robust and convincing means by which it can be examined critically.' He continued to argue

41 Nicolas Bourriaud, *Relational Aesthetics* (Dijon: Les Presses du Réel, 2002, orig. 1997), p. 14.

42 An important criticism of Bourriaud's position is that he narrowly seems to prioritise the aesthetic in his appreciation of such human interactions with their social systems. Such objections have been voiced articulately by Claire Bishop who has questioned the utopian need in Bourriaud's theory for a 'unified subject as a pre-requisite for community-as-togetherness', as unrealistic given the 'divided and incomplete subject of today'. Claire Bishop, 'Antagonism and Relational Aesthetics', *October* (110, Fall 2004), pp. 51–79.

that a potential problem facing discourse concerning so-called new media art was how it had been contextualised and historicised. In other words, as Gere explained, the point was 'not that there is no critical discourse, but rather that it remains the preserve of those involved, with little or no connection or engagement with outsiders'.[43]

Gere's claim is one that art historians should take very seriously. However, this requires that a central problem needs be overcome. This problem concerns the traditional art historical preoccupation with specificity of media and the question of how it may be reconciled with the proliferation of differing media employed in new media art. W.J.T. Mitchell diagnosed the problem in the following terms:

> In the field of art history, with its obsessive concern for the materiality and 'specificity' of media, the supposedly 'dematerialized' realm of virtual and digital media, as well as the whole sphere of mass media, are commonly seen either as beyond the pale or as a threatening invader, gathering at the gates of the aesthetic and artistic citadel.[44]

However, given the wide variety of practices characterised by the systems aesthetic the art historian can take nothing for granted. This is especially the case in new media art which Lev Manovich argues, in *The Language of New Media*, is characterised by 'variability':

> A new media object is not something fixed once and for all, but something, that can exist in different, potentially infinite versions. This is another consequence of the numerical coding of media … and the modular structure of a media object.… Instead of identical copies [of what Manovich calls 'old media'] a new media object typically gives rise to many different versions. And rather than being created completely.[45]

Whilst the problems of critical purchase are undoubtedly foregrounded by new media art, they are not unique to it. However, if there is to be a convincing theoretical model for new media art, then it should not be tied to a specifically narrow set of artistic practices be-

43 Charlie Gere, "New Media Art", *The Art Book*, vol. 12, 2 (2005), pp. 6–8.

44 W.J.T. Mitchell, *What Do Pictures Want?* (Chicago: University of Chicago Press, 2005), p. 205.

45 Lev Manovich, *The Language of New Media* (Cambridge, MA: MIT Press, 2001), p. 36.

cause this will rapidly open it to charges of anachronism or lack of relevance. Today's new media quickly become defunct, and themselves become objects of nostalgia and aestheticisation. For example, the pioneering computer art of Ben Laposky and his *Electronic Abstractions* (from 1956 onwards) is a world away from the complex and aesthetically involving environments of recent popular video and computer games such as Tetsuya Mizuguchi's *Rez* (for Sony Playstation and Sega Dreamcast) or Valve Software's *Half-Life* I & II. Equally, given that they are all constituted by particular historical and technological social variants, the early video art of Bruce Nauman and Vito Acconci from the early 1970s has a clearly different aesthetic to that of the epic film projects of Matthew Barney's *Cremaster Cycle.* W.J.T. Mitchell discussed some of these problems with anachronism in his *What Do Pictures Want?* (2005). The first two of the 'Ten Thesis on media' he outlines there are: '1 – Media are a modern invention that has been around since the beginning (sic.). 2 – The shock of new media is as old as the hills.'[46] Today's new media, in other words, will always be tommorrows old media.

Systems aesthetics and the vocabularies of systems theory provide, I argue, the basis for such an 'appropriately robust and convincing' theory of new media art. They do so by expanding the discourse on new media beyond a discussion of a narrow set of art practices corresponding to a limited set of media into a discussion about systems art more generally. Systems theory and systems aesthetics thus employs the idea of system as medium to inscribe a coherency into what would otherwise seem to be utterly disparate works. This opens these works up to art historical analysis and provides continuity with historical precedents, which may, in the first instance, appear materially incomparable. My two closing examples demonstrate this.

Jeffrey Shaw is best known for his interactive multi-media environments. In *Place–Ruhr* (Computer graphic/ video installation, 2000) the viewer negotiates eleven live-action and three-dimensional virtual environments from the industrial German region of Ruhrgebiet. The viewer participates in the work from a platform and an interface centrally located within the cylinder of the installation projection screens.

46 Mitchell, *What Do Pictures Want?*, p. 211.

From this platform the viewer controls both their spatial relationship to the screens and their progress through the environments that are realised as a sequence of cylinders. Upon entering these panoramic cylinders the environment is cinematically generated around the spectator in relation to them. The installation contains further elements of interaction by using sounds made by the participant to trigger projected words that move through the environment in paths dictated by the movement of the viewer. Mark Hansen describes this process of interaction in such computer-aided works as part of an embodied, phenomenological experience that:

> ... specifically invest the body as the site of a bodily, but also an 'intellectual' event. In these works, the body, rather than being assimilated into the deframed image-space, stands over against a now vitualized image-space and thereby acquires a more fundamental role as the source of the actualization of images. If the corporeal and intellectual processing it performs still functions to 'give-body' to the image, it does so by not lending its physical, extended volume as a three-dimensional screen for the image but rather by creating an image-event out of its own embodied processing of information.[47]

Shaw creates multi-media systems which the embodied participant actively reconfigures with each immersive interaction. It is thus an example of Burnham's model of systems aesthetics in which aesthetic sensations are produced without an intervening object and in which the 'viewer' fully participates in the production of the system.

My second example is provided by Pierre Huyghe and Phillipe Parreno's collaborative project *No Ghost Just a Shell* (1999–2002), which is another contemporary expression of systems aesthetics. For the project the artists acquired the digital files and copyright for a generic, minor, female Manga character called Annlee. (it cost $400, about 46,000 Japanese Yen). The character was an 'avatar'; that is sufficiently anonymous to be used in any story and not expected to play anything other than a minor role. She was a prepubescent girl with wide eyes, blue hair and little in the way of specific features. Huyghe and Parreno commissioned a series of twelve works from different artists for an exhibition at the Zurich Kunsthalle and the San

47 Mark Hansen, *New Philosophy for New Media* (Cambridge, MA: MIT Press, 2004), p. 60.

Francisco MOMA (2002/03). The works included animations, paintings, posters, books, neon works and sculptures. Thus AnnLee is both a 'shell' for a variety of different manifestations and the medium for those manifestations. For example Rirkrit Tiravanija's *Untitled (Even Electric Sheep Can Dream)* (2002) is an eight-hour digital animation in which the character Annlee reads the entire text of Philip K. Dick's *Do Androids Dream of Electric Sheep*; Liam Gillick's *Annlee You Proposes* (2001) is a digital film in which the character explores her environment and M/M produced silkscreen posters (2002).

No Ghost Just a Shell provides just one example of a post medium practice that does not cohere with a specific material medium. Each of the different manifestations of it are a different form of Annlee. Crucially, each form of Annlee is materially very different from other forms and there is no material specificity to Annlee as a medium. Annlee is a set of conditions that may take a variety of material forms. Thus Annlee is a system that provides the possibility of forms within it (or a 'shell' to be filled); but also there would be no Annlee without the various forms by which it can be observed. It is my argument that Annlee is both a system and a medium. That is Annlee is a set of elements integrated with one another to such an extent that they form a recognisable and coherent whole that performs some recognisable function. And Annlee is also recognisable as the medium of the work as she is a set of possibilities that allow for different and specific forms to emerge. *No Ghost Just a Shell* is thus exemplary of the move from object to system in the systems aesthetic; a move that, crucially, may have very different artistic outcomes (as illustrated by *Spiral Jetty* in the next chapter).

Chapter 4: The Work of Art as System

NO CHAOS DAMN IT
Jackson Pollock[1]

Introduction

What does it mean to talk about artworks in terms of systems? This chapter uses artworks by two specific artists (Jackson Pollock and Robert Smithson) to explore the application of systems theory to art. It is my argument that works of art are not closed systems with specific representational content. Instead they can be observed via the systems aesthetic as situated within complex and distributed (not closed) systems.

When art historians approach a work of art the key question they traditionally ask is what is represented? There are a number of art-historical strategies for answering this question ranging from considerations of style and iconography to other cultural and historical analyses. Using systems theory as an alternative strategy, a major challenge is, therefore, to explain in what sense systems can be said to be represented by works of art. In this chapter I use the abstract work of Jackson Pollock as my main example in tackling this challenge. In what follows I will discuss several different (and I argue ultimately

1 In 1950, *Time* magazine printed parts of an essay by Bruno Alfieri, previously published in Italy, which used the word 'Chaos' to describe the paintings of Jackson Pollock. Unable to let the criticism pass, Pollock sent *Time* a telegram that began, 'NO CHAOS DAMN IT'. From James Coddington, 'No Chaos Damn It', in Kirk Varnedoe and Pepe Karmel (eds.), *Jackson Pollock: New Approaches* (MOMA, 1999), pp. 101 ff. Coddington discussed the order of Pollock's paintings in relation to his method but does not refer to chaos-theory, fractal analysis or systems theory in doing so.

unsuccessful) ways in which one could show that his works represent (closed) systems. I do this in order to demonstrate a more appropriate application of systems theory and the systems aesthetic to works of art in general; namely one that observes art as situated within complex systems. I close with a discussion of Robert Smithson's work as an example of such an approach.

I have chosen to focus mainly on the work of Jackson Pollock for three reasons.

First, as I will discuss below, Pollock's works have been subject to an unsatisfactory systems analysis, which claimed to find fractal patterns in his paintings. Visual evidence of complex and systemic patterns was identified, yet in a manner which, I argue, did not do justice to the radical implications of a systems theoretical approach.

Second, the encounter with abstract, non-representational art such as Jackson Pollock's *Autumn Rhythm: Number 30, 1950* (illus. 6) highlights the problems inherent in seeking to establish what artworks represent. This is because in the apparent absence both of a clear distinction between iconographic figure and ground and of a recognisable schema of signs and symbols, the ambivalence and contingency of the representational content of artworks is thrown into relief. If, on the other hand, we approach the work in observational (rather than representational) terms, we recognise that what is observed is contingent both upon the process of observation and upon the complex and distributed systems within which that observation takes place.

Third, the ambivalence of representation in art such as Pollock's reflects how an abstract work of art is itself a figure observed against the ground of art historical discourse. This is to say (as discussed in detail in the final chapter) that in the discursive systems of art the identity of the art work itself is fluid rather than static, ambiguous rather than definitive, and abstract rather than concrete. In other words when dealing with abstract art the art historian is confronted with a dual challenge in ascertaining what it represents. This is a challenge that, whilst not specific to it, is certainly fore-grounded by abstract art. This dual challenge consists of having to construct meaningful content in the face of a work of art that not only lacks a clear meaning but also fails to suggest any single and obvious strategy for determining any such meaning.

In the first and main part of the chapter I pose the question: What is the representational content of Jackson Pollock's *Autumn Rhythm: Number 30, 1950*? In response to this question I survey three possible, and unsatisfactory, answers that are based upon a fractal (and thus systemic) analysis of the image.[2] These answers are couched in terms of:

1. Nature/mimesis/representation – by which it is suggested that the image, by virtue of representing fractal forms, can be said to represent nature in some sense.

2. Fractals – by which it is suggested that the image can be conceived in formalist terms as representations of their fractal patterns.

3. Process – by which it is suggested that image can be described as an index of the chaotic process of its creation.

I argue that none of these suggested answers are a satisfactory application of the systems theoretical approach because they still treat the work of art as a self-contained representative entity. In doing so, they do not extend their attentions beyond the artwork to the systems in which it is embedded. They do not, therefore acknowledge the radical implications of the concept of distributed representation within complex systems and thus fail to recognise how distributed representation frames observation.

Finally, I demonstrate an alternative to this restricted and ultimately inappropriate application of systems theory. I turn to Robert Smithson's *Spiral Jetty* (1970) in order show how it can be observed, as I argue more appropriately, via a concept of complex, distributed systems.

2 As these observations concern self-similar acts of representation being observed at different scales they allegorically mirror the fractal form which they invoke.

1. What is the representational content of Jackson Pollock's *Autumn Rhythm: Number 30, 1950*?

The numerous questions that have been asked about Jackson Pollock's artworks represent reflect the cluster of disparate analytical critiques art historians have at their disposal. Marxist, feminist, queer, psychoanalytic, iconographic, iconologic and formalist strategies all advanced different views on various aspects of his work in response to different questions they have asked of it.

Yet in 1999, in two relatively unassuming papers that appeared in *Physics World* and *Nature*, two physicists made a quietly astonishing art historical claim. Richard Taylor, Adam Micolich and David Jonas stated that through the application of 'scientific objectivity' they could determine the 'fundamental content' of Jackson Pollock's major work: the so-called 'all-over paintings' he produced between 1940 and 1952.

To determine this 'fundamental content' Taylor, Micolich and Jonas' method of objectivity required an analysis of the particular mathematical patterns found in the paintings. It was shown that patterns within the work satisfied the formal descriptions of fractals (in other words, as is discussed below, fractals are patterns that correspond to complex systems). By finding such patterns in Pollock's work Taylor et al. thought they had proven that Pollock's work was a complex visual system. Their approach thus demonstrates one way of applying the concept of system to artworks. However, this way is ultimately unsatisfactory; both for its formalist pre-occupations and for its reliance upon a traditional conception of representation.

1.1 What is a fractal?

Stated simply, a fractal is a self-similar pattern.[3] The pioneer of fractal geometry, Benoit Mandelbrot, described fractals as patterns that are the same at different levels of magnification.

Broadly speaking, mathematical and natural fractals are shapes whose roughness and fragmentation neither tend to vanish, nor fluctuate up and down, but remain essentially unchanged as one zooms in continually and examination is refined. Hence the structure of every piece holds the key to the whole structure.[4]

Importantly, a fractal need not necessarily look the same at different scales.[5] Despite this, it can be described using the same rules.[6] That is, it demonstrates *statistical self-similarity* (Voss);[7] the same, simple mathematical rule describes the macro and micro elements of the particular complex structure.

The von Koch Curve (described by Helge von Koch in 1904) is one of simplest ways of imagining a fractal (illus. 7). It is a pattern that follows the simple rule of repetition of a simple triangle shape. Its construction is described by Richard Voss thus:

> A simple line segment is divided into thirds and the middle segment is replaced by two equal segments forming part of an equilateral triangle. At the next stage in the construction each of these four segments is replaced by four new segments with length 1/3 of their parent according to the original pattern. This procedure, repeated over and over, yields the beautiful von Koch curve.[8]

Therefore, by the continued iteration of the initial rule of construction, an image emerges which will be infinitely complex in so far

3 The literature on fractals is vast. The best introduction to the often baffling theory and application of fractal geometry is found in James Gleick, *Chaos* (London: Vintage, 1998). The classic work on fractals is Benoit Mandelbrot, *The Fractal Geometry of Nature* (San Franscisco: WH Freeman and Company, 1977), which strikes a balance between accessible descriptions of the implications of the theory and obscure mathematics. More recent discussions on the applied use of fractal geometry include: M. Fleischmann, D.J. Tildesley, R.C. Ball (eds.), *Fractals in the Natural Sciences* (Princeton: Princeton University Press, 1999); Heinz-Otto Peitgen and Dietmar Saupe (eds.), *The Science of Fractal Images* (New York: Springer, 1988); Heinz-Otto Peitgen and Dietmar Richter, *The Beauty of Fractals* (New York: Springer, 1986).

as it will obey this simple rule regardless of the scale of magnification at which it is viewed. The edge of the Koch curve, comprised as it is of ever diminishing triangles, is a loop which tightens in upon itself to a greater degree the higher the magnification. Further, just as the detail of the edge of the image will continue to increase as the scale of magnification of the image increases to infinity, so too the length of the line around the space enclosed by the edge of the fractal is infinite. Therefore if we take the initial triangle from which the curve is extrapolated, as having sides of length 1 then, as Gleick explained:

> The length of the boundary is 3 x 4/3 x 4/3 x 4/3…. – infinity. Yet the area remains less than the area of a circle drawn around the original triangle.[9]

Gleick's description highlights a key paradox of the fractal: the incongruity between the area of the fractal, which can be shown to be less than the area of a circle drawn around the whole image, and the possibility that this finite area is contained within an infinitely long edge.

Fractals are most likely to appear in complex and dynamical systems which are poised in a state of non-equilibrium; that is, within systems described by Broomhead and Jones as systems whose: 'dynamical behaviour is governed by non-linear evolution equations.'[10] This includes such dynamically chaotic systems such as those which

4 B. Mandelbrot, 'Fractal Geometry: what is it, and what does it do?', in Fleischman et al., *Fractals in the Natural Sciences*, pp. 3–16.

5 In the terms of Chaos Theory, self-similarity is expressed in terms of patterns having the same mathematical description rather than being visually identical.

6 'The key to fractal geometry's effectiveness resides in a surprising discovery that the author has made, largely thanks to computer graphics. The algorithms that generate the other fractals are typically so extraordinarily short, as to look positively dumb. This means that they must be called "simple". Their fractal outputs, to the contrary, often appear to involve structures of great richness' Mandelbrot, 'Fractal Geometry', p. 6.

7 Richard Voss, 'Fractals in Nature: From Characterization to Simulation', in Peitgen/Saupe, *The Science of Fractal Images*, p. 30.

8 Ibid., p. 26.

9 Gleick, *Chaos*, p. 99.

10 Broomhead and Jones, 'Time-Series Analysis', in Fleischmann et al., *Fractals in the Natural Sciences*, p. 103.

have been observed in the natural world, such as those seen in neuron activity (in conscious systems), chemical reactions, fluid flows and economic systems, all of which have been modelled effectively using the methods of fractal geometry.[11] Further examples of this are provided by Gleick who described how problems in the mapping of human anatomy have been solved through the application of fractals. Examples that illustrate the wide variety of systems displaying fractal patterns include: the urinary collecting system,[12] the edges of clouds and the economic system of 20th century cotton prices as analysed by Mandelbrot.[13]

In short, despite the visual complexity of fractal patterns they conform to the simple rules of their construction regardless of the scale of magnification at which they are viewed. The individual complex pattern is underpinned by a structure that can be explained by a few, relatively simple rules. For the cited study of Pollock's work this meant that if a similar type of pattern could be discovered in his paintings then it could be concluded that they represented complex systems.

1.2. Pollock and Fractals

To demonstrate the existence of fractals within Pollock's paintings Jonas et al. analysed several paintings. The chosen paintings were

11 See especially P. Fischer and William J. Smith (eds), *Chaos, Fractals and Dynamics* (New York: Marcel Dekker Inc., 1985), and Peitgen/Saupe, *The Science of Fractal Images*.

12 Gleick, *Chaos*, p. 109.

13 'Indeed when Mandelbrot sifted [Houthkakker's] cotton-price data through IBM's computers, he found the astonishing results he was seeking. The numbers that produced aberrations from the point of view of normal distribution [and the classical economical model of supply and demand] produced symmetry from the point of view of scaling. Each particular price change was random and unpredictable. But the sequence of changes was independent of scale; curves for daily prices and monthly price changes matched perfectly. Incredibly, analyzed Mandelbrot's way [according to fractal patterns] the degree of variation remained constant over a tumultuous sixty-year period that saw two World-Wars and a depression.' Ibid., p. 86.

scanned into a computer, overlaid with a computer generated grid of identical squares and analysed at various levels of magnification from squares of length 0.8mm (the detail of the very finest paintwork) to squares of around 2.5 meters (a square containing the whole canvas).[14]

Jonas et al. found that in several of the paintings they investigated there were two co-existent fractal patterns: the first pattern corresponding to the fine paintwork at a scale between 1mm and 5cm; and the second pattern corresponding to the broad paintwork at scale between 5cm and 2.5 m.[15] Both patterns can be expressed mathematically as what is called the 'fractal dimension' of the pattern. Fractal dimensionality is obscure. It is the method Mandelbrot used to describe the relationship the viewer has to the object with regards to the scaling, or the level of detail at which the object is analysed. It expresses the level of detail at which the object is self-similar in relation to the observer's level of scrutiny.[16]

Contrary to the three dimensions of Euclidean geometry (height, depth and breadth) with which we are intuitively familiar, fractal dimensions need not be an integer. Fractal dimension is a way of expressing a level of detail of a self-similar shape in relation to the observer's level of scrutiny. Thus the fractal dimension, or D, provides an expression of the level of detail of that shape, or, as Voss has

14 Richard Taylor, Adam Micolich, David Jonas, 'Fractal Expressionism', *Physics World*, vol. 12, 10 (1999), pp. 25–28.

15 'This fractal analysis reveals two distinct D values occurring over the ranges 1mm L < 5cm and 5 cm < L 2.5 M.' from Taylor, Micolich, Jonas, 'Fractal Analysis of Pollock's Drip Paintings', *Nature*, vol. 399, 3 (1999), p. 422.

16 Peitgen/Saupe, *The Science of Fractal Image*s, p. 29. Gleick describes this concept of fractal dimensions thus: 'What is the dimension of a ball of twine? Mandelbrot answered, it depends on your point of view. From a great distance the ball is no more than a point, with zero dimensions. From closer the ball is seen to fill spherical space, taking up three dimensions. From closer still, the twine comes into view, and the object becomes effectively one-dimensional, though the one dimension is certainly tangled around itself in a way that makes use of three dimensional space. The notion of how many numbers it takes to specify a point remains useful. From far away it takes none – the point is all there is. From closer it takes three. From closer still, one is enough – any given position along the length of twine is unique, whether the twine is stretched out or tangled up in a ball.' Gleick, *Chaos*, p. 95.

explained: 'the fractal dimension D, thus, provides a quantitative measure of the wiggliness of the curves.'[17] Mandelbrot argued that for certain shapes their dimensionality is not best expressed in integers; rather in fractions. They are, in other words, not clearly defined by one scale but instead have a certain, yet characteristic, irregularity that remains constant over different scales. This is the regular irregularity of the self-similar and infinite patterns suggested by fractals.[18]

In short, the fractal dimension, or D, provides an expression of the level of detail of that shape.

Thus, for example, *Blue Poles Number 11* (1952) was found to demonstrate two fractal patterns, the first of which was the close detail fractal, or D_d – having a value 1.72, with the larger scale pattern, or D_L having a value that was stated to be 'quite close to 2'. In all the paintings analysed, the second, large scale, fractal was found to have a value close to 2, and the following values were found for the small-scale fractal pattern, D_d – *Untitled: Composition With Pouring II* (1943) – had a D_d value 'close to 1', *Number 14, 1948* (illus. 8) had a D_d value of 1.45 and *Autumn Rhythm: Number 30, 1950* a D_d value of 1.67.

The question at this point should be: but what does this prove? It is worth recalling that Jonas et al. had claimed that through the application of 'scientific objectivity' they could achieve a key objective of art historical investigation; namely the discovery of the 'fundamental content' of the images. Their conclusion was that the fundamental objective representational content of *Autumn Rhythm: Number 30, 1950* was the two fractal dimensions: (1) a close detail fractal, or D_d of value *1.67*; (2) a larger scale fractal pattern, or D_L of a value that was stated to be 'quite close to 2'.

17 Peitgen/Saupe, *The Science of Fractal Images*, p. 29.

18 'The large amount of repeating structure within a fractal pattern causes it to occupy more space than a smooth one-dimensional line, but not to the extent of completely filling the two dimensional plane.' Taylor et al., 'Fractal Expressionism', pp. 25–28.

However, it would seem that if the fundamental content of *Autumn Rhythm: Number 30, 1950*, is 1.67 and nearly 2, then the questions we are asking of it need some further investigation and refinement. The question as to what is represented by such fractal patterns needs some further unpacking, as follows below.

1.3 Nature/Mimesis/Representation

First it could be argued that existence of fractal patterns in Pollock's all-over canvases provide evidence of structures in the work that are mimetic of naturally occurring forms. In this sense, then, they can be understood as representational. This is because they correspond to naturally occurring fractal patterns. As has already been observed, fractal patterns occur naturally in nature (for example in bracken fronds and capillaries in the lungs). Jonas et al. also note that natural fractal patterns such as coastlines and lightning display a fractal dimension of between 1.25 and 3. Further they point out that systems that have two fractal dimensions, like Pollock's paintings, are also naturally occurring. The examples they provide are 'trees and bronchial systems'. Thus, by illustrating fractal patterns and demonstrating such dynamic, systemic patterns as are naturally occurring, the paintings could be seen to be mimetically representative of naturally occurring dynamic systems.

The dynamism of the natural world is seen then as re-presented within the complex artistic systems of the paintings. This certainly supports a popular reading of Pollock's work, namely that it is somehow natural; a reading which Pollock himself evoked with titles such as *Autumn Rhythm*. For example, Martin Gardner suggested that 'photographers with a keen sense of aesthetics find it easy to take photomicrographs of natural patterns that are almost indistinguishable from abstract expressionist art.'[19] Other art historical readings have suggested that the patterns of seaweed Pollock saw whilst mussel farming on

19 Larry Short, 'The Aesthetic Value of Fractal Images', *British Journal of Aesthetics*, 31:4 (1991), p. 351.

the beach on Long Island have worked their way into the work via such paintings as *Blue Poles*.[20]

1.4 The Fractal Itself

The second interpretation is a more rigidly formalist reading of the fractal patterns. It suggests that what is represented is the fractal itself. This question takes the painting as a closed system of representation gesturing toward the conceptual object of a mathematical model; the icon of which is the fractal pattern.

This second approach thus brings the response within the sphere of the claims which have been made for other fractal images. If Pollock's paintings are fractals, then it evidently follows that what has been claimed for fractal images may also be claimed for them. Fractal images provided as representations of mathematical models have been applied to aesthetic analysis, like artworks. And it has been suggested that in their graphical representation of the non-representable infinite regress of self-similarity, they might display aesthetic qualities.

Certainly this is how Mandelbrot conceived of his fractal geometry, as a 'world of pure plastic beauty unsuspected till now'.[21] In this respect then, fractals arguable provide an ultra-modernist artwork; an expression of Mondrian's 'pure plastic art'. Mondrian's definition of his 'pure plastic art' would indeed seem to fulfil the criteria of a fractal in so far as:

> The important task of all art, then, is to destroy the static equilibrium by establishing a dynamic one. Non-figurative art demands an attempt of what is a consequence of this task, the destruction of particular form and the construction of a rhythm of mutual relations, of mutual forms of free lines.[22]

20 Steven Naifeh and Gregory White Smith, *Jackson Pollock: An American Saga* (New York: Clarkson N. Potter, Inc., 1989).

21 Mandelbrot, *The Fractal Geometry of Nature*, p. 4.

22 Piet Mondrian, 'Pure Art and Pure Plastic Art' (1936), in Herschel Chipp (ed.), *Theories of Modern Art* (Berkeley: University of California Press, 1968), pp. 349 ff.

Following this line of argument, the self-similar, fractal patterns of *Autumn Rhythm: Number 30, 1950*, as do other visual representations of fractals, such as the image of a Julia Set (illus. 9), disclose a visual system which is in a state of dynamic non-equilibrium and which, by virtue of its scaling self-similarity, is a 'construction of a rhythm of mutual relations, of mutual forms of free lines'. Couched in these terms, fractals (Pollock's included) are exemplary of modernist 'pure plastic art'.

Developing this aesthetic analysis further, in an article which appeared in *Modern Painters*, P.W. Atkins claimed that images based upon fractals fulfilled the description of being a 'hypermodern art'. He argued that in its synthesis of aesthetic and scientific concerns the fractal could provide the conceptual basis for an art suitable for our technologically and scientifically advanced society. From his starting point that 'it is plausible that a rigorously scientific basis of aesthetics will one day be available', Atkins proposed that aesthetic response results from the 'exposure' of the 'appropriately resonant circuitry'[23] of the brain to the aesthetic object that is the focus of the aesthetic attention. Whilst Akins admits such a determinist project is beyond the present state of knowledge in biology and physics knowledge, he is convinced that such a model of aesthetic response, reminiscent of the 'free-play' of reason and imagination in Kantian aesthetics, theoretically can be conceived in materialist terms as being a function of the organic system of consciousness. In conclusion Atkins goes on to state that because fractals are mathematically, intellectually and aesthetically both engaging and pleasing they are 'beautiful' images. This is because, he argued, at 'one level the warmth of our response to it must lie in the dizzying richness of the image. A fractal image is an image without end.' Fractals are, for Atkins, in their simple, yet theoretical infinite complexity 'deep, inexhaustible wells of beauty'.[24]

Following this line of thought fractals Pollock's paintings such as *Autumn Rhythm: Number 30, 1950* can also be considered to fulfil the

23 P.W. Atkins, 'The Rose, The Lion, and the Ultimate Oyster', *Modern Painters*, vol.2, 4 (Winter 1989/90), pp. 50–55.

24 A similar line of argument is pursued in Short, 'The Aesthetic Value of Fractal Images', pp. 342 ff.

strictly formal (and mathematical) conditions of being 'deep, inexhaustible wells of beauty', and are, therefore, a fulfilment of the criteria of a hypermodern art which Atkins talked about. The logical conclusion of this would seem to be that if both fractals are inherently aesthetic, and the canvases are inherently fractal, then it is possible to provide an algorithm for beauty in art. At the heart of this take on fractals is the suggestion that there can be a mathematically formalised set of formal procedures (algorithms) lying at the core of the visual systems of pure plastic, hypermodern art.

This plays both like a parody of Greenberg's formalism, which explains the beauty of Pollock's work by the inherent form of the underlying structure, and like an inversion of Kant's argument, according to which mathematical forms can't be beautiful precisely because of their self-similarity.

1.5 An index of process

The third interpretation of the fractal patterns in the paintings relates to their patterns as an index of the process of their creation. As Rosenberg suggested in 'The American Action Painters' (1952) one way of looking at the complex visual system of Pollock's paintings is that what is being represented on the canvas is 'not a picture but an event'.[25] The patterns are thus representative of Pollock's movements around the canvas and hence serve as a visual record not only of his own movements but also of the application of the paint itself whether it was placed by Pollock or let to fall by its own devices. That this index of creation and of the application of the paint is fractal suggests that the way in which the paint fell was chaotic and hence dynamic in its nature in the same way as other natural processes that can be modelled using the visual analogue of the fractal.[26]

25 Harold Rosenberg, 'American Action Painters', in Charles Harrison and Paul Wood, *Art in Theory 1900–1990* (Oxford: Blackwell, 1991).

26 Whirlpools and sand piles are two such examples discussed by Mark Buchanan, in *Ubiquity* (London: Weidenfeld and Nicolson, 2000).

As already noted, in the paintings analysed by Jonas et al. two fractal dimensions were identified, corresponding to two different types of patterns suggesting two processes of construction. Jonas et al. acknowledged this and suggested that the two fractal dimensions represented the following:

First: an all-over, large-scale pattern which could be seen as the visual scaffolding of the paintings. This pattern corresponds to Pollock's movement around the canvas. It was identified as having a fractal dimension of 'nearly 2'. Significantly this figure is the same for the larger pattern in all of the canvases which were analysed. This suggests that a similar process was being employed each time. The pattern is the anchor layer of the painting, upon which the complete work is built upon, around and within.

Second: a smaller scale fractal pattern which fills in the detail of the scaffolding of the anchor layer. This pattern is an index of a finer, more detailed working approach which concentrates upon particular areas of the canvas at any one time (as opposed to the anchor layer which involves working on the canvas as a whole). It was suggested that this pattern is the *detail* layer of the painting. Taylor, Jonas and Micholich maintained that this smaller scale pattern represents Pollock's dripping technique which was 'refined' between the years 1943 – 1952. And they concluded:

> Our analysis shows that Pollock refined his dripping technique: the fractal dimensions increased steadily through the years from close to 1 in 1943 to 1.72 in 1952.[27]

They claimed that this increase in the fractal dimension, simply meaning that the pattern became more expansive, indicated that the pattern of this detail layer became denser in Pollock's later abstract works. Their method therefore must, in their view, be considered a legitimate art historical technique: 'fractal analysis could be used as a quantitative, objective technique both to validate and date Pollock's drip paintings.'[28]

27 Taylor et al., 'Fractal Analysis', p. 422.
28 Ibid.

The claim that these two fractal patterns inherent in Pollock's visual systems correspond to different working practice can be validated by looking at accounts of Pollock's method of working on the abstract canvases. These accounts corroborate the argument that the patterns are visual indicators of the dynamic events of their creation. First hand accounts of Pollock at work by Goodnought, Namuth (who filmed Pollock) and Karmel[29] (who has produced an exhaustive study based upon Namuth's films and other accounts of Pollock at work) demonstrate, on one hand, Pollock's broad approach to the whole canvas (corresponding to the fractal of the anchor layers); and, on the other hand, his more detailed close working of the canvas (corresponding to fractal the detail layers).

The accounts we have of the processes involved in creating *Autumn Rhythm: Number 30, 1950* (illus. 10) demonstrate this. In an article, which first appeared in *Art News* in 1951, Goodnought described that at the start of the process Pollock began with a blank, clean canvas, which was worked in a period of intense and absorbing activity (about 2 hours) at the end of which Pollock was 'completely exhausted'.

> After a while he took a can of black enamel... and a stubby brush which he dipped into the paint and then began to move his arm rhythmically about, letting the paint fall in a variety of movements on the surface. At times he would crouch, holding the brush close to the canvas, and again he would stand and move around it or step on it to reach to the middle. Within half an hour the entire surface had taken on an activity of weaving rhythms... As he continued, still with black, going back over former areas, rhythms were intensified with counteracting movements.[30]

Then, Goodnought related, Pollock would spend two weeks becoming accustomed to the painting, after which he would sporadically apply different layers of paint to the canvas in a more deliberate fashion. The third and final working of the canvas, some weeks after the initial act of construction, was described as being: 'Slow and delib-

29 All quoted in Pepe Karmel, 'Pollock at work: The Films and Photographs of Hans Namuth', Kirk Varnedoe with Pepe Karmel, *Jackson Pollock: New Approaches* (New York: The Museum of Modern Art, 1998).

30 Ibid.

erate. The design had become *exceedingly complex* and had to be brought to a state of complete organisation.'[31]

Whilst Karmel discussed inconsistencies with Goodnought's account, it is clear nonetheless that Pollock did work at different levels of intensity at different periods of time until the canvas as a whole was complete. Goodnought's identification of three types of activity (the initial work, the complementary changes and the final, close detailing) did, therefore, correspond with Taylor, Jonas and Micholich's argument that the different fractal dimensions represent different working processes; on the one hand a canvas-wide action, and on the other a closer detailing of the pattern. Karmel also provided evidence of this taken from his close scrutiny of the two films Namuth made of Pollock at work. To demonstrate this Karmel used some of Namuth's colour film out-takes from the filming of Pollock working on a now lost painting made on red canvas. Karmel suggested that this footage refutes the idea that the paint was applied rapidly at random, but instead:

> Step by step evolution of the painting from bare canvas to *a complex web*, [demonstrating Pollock's] combination of kinetic freedom and formal control.[32]

As Karmel also observed with reference to one sequence of film:

> A close look at several frames [of the film] sequence underscores the fact that the *splats are produced by a different hand motion than the initial curve*s. After taking the brush from the paint can, Pollock lifts his hand and bends the brush inward, toward his body; he then snaps his forearm downward, finishing with his palm down rather than up. It is the same gesture of wind-up and release visible in some of the still photographs of Pollock working on *Autumn Rhythm*.[33]

Such a working method was also seen in the section of the second film where Pollock painted onto glass, with the camera beneath.

There are two important parts of this statement. Firstly the observation that there is visual evidence of Pollock's differing techniques which again corroborates the argument that the two fractal patterns

31 Ibid., p. 95; emphasis added.
32 Ibid.
33 Ibid., p. 114.

correspond to two differing types of activity. And secondly there is the observation that Pollock used the same type of motions between canvases (in this case the lost, red canvas painting and *Autumn Rhythm: Number 30, 1950* – which was subjected to fractal analysis.) This further confirms the argument that there is a fractal similarity shared between (a selection of) the canvases.

1.6 Problems with the fractal analysis

Thus far I have offered three suggestions as to what the fractal patterns in Pollock's paintings might represent. Yet, as stated from the outset, such analyses are unsatisfactory. All three of these interpretations are, at heart, formalist. They deal with the specific formal qualities of the paintings; and the formal analysis reveals fractal patterns. There are two main problems with the method proposed by Jonas et al.

First there is a problem of inversion. The scientists are using fractals to represent Pollock's paintings whilst claiming that the paintings represent fractals. We know specifically that Pollock did not attempt to represent fractals. Instead the fractal patterns have been used as an interpretive framework by which to re-present the paintings; just as you can use fractal patterns to describe and re-present Coastlines and bracken fronds. Yet we don't think of coastlines and bracken fronds as representing anything.

Second there is a problem of appropriateness. To treat a painting as if it were a coastline is to merely deliver an interpretation of its mathematical structure without actually saying anything particularly interesting about it. Such a strategy leads, ultimately, to historically inert claims about the paintings based on numerical values.

A more satisfactory reading of Pollock's work will move to a different scale of observation. An effective account of any work of art must extend its focus beyond the singular work itself to the historical and political contexts of its production and reception that is to the complex and distributed social systems within which it is situated.

The importance of Pollock's all-over paintings in the development of the avant-garde lies in his dismantling of the traditional basis of pictorial representation; namely the distinction between figure and ground. As Clement Greenberg argued in 'The Crisis of the Easel Picture' (1948), Pollock's importance as a painter is that he suggested 'a way beyond the easel, beyond the mobile, framed picture',[34] Greenberg's reading of Pollock's canvases as flat planes, or all-over pictorial fields means that the perceptual difference between figure and ground is negated. There are two dominant readings of this.

First in Greenbergian terms this places Pollock at the end of a developmental narrative of modernist painting that ends with the move from representation to the articulation of the fundamental flatness of the medium of painting. However, this was a point beyond which Modernist painting historically did not develop.

Second the effacement of figure by ground also places Pollock within the narrative of the 'informe', formless or anti-form where, as Krauss has argued, Pollock's dismantling of the conventions of the easel picture was subsequently investigated in further media by Robert Morris, Richard Serra and Andy Warhol (amongst others).

The fractal analysis, however re-inscribes traditional pictorial and aesthetic values within Pollock's paintings. It does so by re-inscribing resemblance back into the images by reading the fractal as an identifiable figure against the ground of the pictorial spaces behind it. The re-emergence of the figure/ground distinction thus re-inscribes Pollock within the tradition of representational easel painting rather than investigating the historical conditions by which he heralded its dissolution.[35]

And herein lies a certain irony – namely the point at which Pollock's paintings become formless and chaotic; they become available to fractal analysis. Or, in other words, when figure and ground are effaced in one representative order, they re-appear in another.

This paradox is illustrated further by a particular example from Karmel. Karmel finds an initial figurative order to Pollock's visual

34 Clement Greenberg, 'The Crisis of the Easel Picture' (1948), *Art and Culture* (Boston: Beacon Press, 1965), pp. 154–157.

35 'If I am going over this all-too-familiar- ground, it is to underscore the stakes involved in promoting the idea of Pollock as a draftsman, of deciding to read his line as contour rather than its dissolution', Rosalind Krauss, 'The Crisis of the Easel Picture', Varnedoe/Karmel, *Jackson Pollock*, pp. 155–179.

system. In contrast to Goodnought's account of Pollock's tri-partite working method, Karmel suggests the work has four stages with the first, intense period of activity being split into two also distinct actions. Firstly there was an initial tracing of calligraphic and loosely representational figures (clearly distinct from the ground) followed by the second task of immediate over-painting and obscuring of these figures.

In one example, *Number 27, 1950*, the painting was begun with black lines and a series of loose arcs that are vaguely descriptive of the Picasso-esque motifs, which appear elsewhere in Pollock's work. Karmel has related this inception of the work:

> Watching from above, we see a hand extend a paint-clotted brush a few inches above the bare canvas. A black line appears on the white surface below it. As the brush moves through the air, the line swerves to the left, curves upward, and then loops to the right, forming a pumpkin shaped ovoid. Narrow, then thick, then narrow again, it terminates in a pair of tiny blots connected by a stem; a miniature barbell with one end light and the other dark.....
>
> In the following nine seconds, Pollock adds a series of lines describing the contours of a figure unmistakably similar to the figures in earlier pictures such as *Untitled (Cut Out)* (1948). One line curves upward on the left to form a shoulder then turns down and right to make an arm, terminating in calligraphic fingers. A second line rises from the left corner of the canvas to describe a foot and a leg. A third descends on the right, bending at the knee. A fourth is bent into a foot, shaped like a triangular shirt hanger. Moving right, Pollock begins a second figure, working upward this time instead of down.[36]

Having stated this Karmel argued next that the randomness of Pollock's painting in the second half of this first stage obscures this initial representational aspect of the picture. Through this, he continued:

> This is the point at which the painting is transformed from a collection of independent pictographs into a single all-over composition united by a consistent rhythm of dark and light, thick and thin extending across the surface.[37]

36 Karmel, 'Pollock at Work', p. 107.
37 Ibid.

Thus it is that at the point at which the question, as to what the representational content of paintings is, becomes impossible to answer from one perspective it becomes possible from another. This is because the point of abstraction is the point at which the fractal figure emerges from the ground. The irony is that by virtue of being non-representative, the paintings end up becoming representative again on a different scale.

A more effective way of reading Pollock's work will require that it is situated within the context of the social and interpretive systems by which it is observed. Thus, as discussed above, Pollock's continued importance in development of avant-garde art is a function of the abstraction of his paintings. That is, the paintings are historically significant because they resist comprehension by resisting representation. This means that attempts to find singular definitions of the representative content of the paintings will always be confounded by the works' complexity. And that this complexity reveals the contingency of the multiple positions from which the works may be observed.

2. The Complex, Distributed System of *Spiral Jetty*

Robert Smithson's *Spiral Jetty* (1970) (illus. 11) is an unstable and chaotic work both physically and conceptually. Superficially it is a 1,500 foot long sculpture in the Great Salt Lake in Utah that was submerged in 1972 and which has reappeared at various intervals (most notably in 2002). Its physical intransigence serves as a powerful metaphor for its shifting conceptual identify. It is this ambiguous identity which presents a challenge in determining what is being represented in the work. And there are different conceptual scales at which the work can be observed.

Robert Hobbs gave the following physical description of the sculptural aspect of *Spiral Jetty* as: 'Spiral Jetty, Rozel Point, Great

Salt Lake, Utah, April 1970; Mud, precipitated salt crystals, rocks, water; coil 1500 ft long and 15 ft wide.'[38]

But there is more to *Spiral Jetty* than its physical constitution. *Spiral Jetty* is a complex work. Different aspects of it include:

1. The Sculpture in the Salt Lake, which Smithson completed in 1970;
2. the 35 minute 16mm film which Smithson made;
3. the essay Smithson wrote about the work;
4. the photographs taken by Smithson and Gianfranco Gorgoni of the work and its creation;
5. the sketches Smithson made in preparation;
6. the performance of the work's creation;
7. the subsequent discourse which surrounds the work, and so on, and so on.

It is my argument here that this multiple aspects of the work constitute something like a system of distributed representation; and that *Spiral Jetty* is distributed throughout, and thus represented by, this system.

Smithson himself acknowledged the complexity of the work in his concept of the site/non-site dichotomy; an unstable interconnection of cross-reference which he explored in his work.[39] He defined non-site in the following terms:

> *The non-site an (indoor earthwork)* is a three dimensional logical picture that is *abstract* yet it represents an actual site... To understand this language of sites is to appreciate the metaphor between the syntactical construct and the complex of ideas, letting the former function as a three dimensional picture which doesn't look like a picture... Between the actual site ...and *The Non-Site* itself exists a space of metaphorical significance. It could be that 'travel' in this space is a

38 Robert Hobbs, *Robert Smithson; Sculpture* (Ithaca: Cornell University Press, 1981), p. 191.

39 Smithson produced a series of *non-sites* including *A Non-site, Pine Barrens, New Jersey*, a sculpture of blue aluminium boxes containing sand from the site accompanied by a aerial photograph/map of the site in New Jersey (Winter 1968; Virginia Dwan); and *A Non-site, Franklin, New Jersey*, a sculpture of wood boxes filled with limestone from the site with accompanying map (Summer, 1968; Museum of Contemporary Art; Chicago).

vast metaphor. Everything between the two sites could become physical metaphorical material devoid of natural meanings and realistic assumptions.[40]

Initially it seems that the concept of the 'non-site' which Smithson proposes is merely a replication of the notion of non-distributive representation which is critiqued from a systems theoretical perspective of *distributed* representation. This is because it is a 'logical picture' (normally in sculptural form; incorporating material from the site) that 'represents' the site; thus preserving the structural relationship between signifier and signified upon which non-distributed representation is founded. However it subverts this by the site being interconnected with the non-site; both of which are in a paradoxical, yet symbiotic process of fulfilment and negation. As Smithson said of this negation:

> In the very name 'non-site' you're making a reference to a particular site but that particular site evades itself, or it's incognito. You're on your own. You're groping out there. There's no way to find out what's there. Yet you're directed out there. The location is held in suspense. The non-site itself tends to cancel out the site. Although it's in the physical world, it's not there.[41]

This is to say that even though the *site* is incomplete without its complementary counterpart it is simultaneously negated and effaced by the presence of the non-site (and likewise the non-site is both fulfilled and negated by the site). Smithson talked about the site/non-site in such terms:

> There's a central focus point which is the non-site; the site is the unfocused fringe where your mind loses its boundaries and a sense of the oceanic pervades as it were. I like the idea of quiet catastrophes taking place.... The interesting thing about the site is that, unlike the non-site, it throws you onto the fringes. In other words, there's nothing to grasp onto except the cinders and there's no way of focusing on a particular place. One might even say that the place has absconded or been lost. This is a map that will take you somewhere, but when you

40 R. Smithson, 'A Provisional Theory of Non-sites' (1968), in Jack D. Flam (ed.), *Robert Smithson: The Collected Writings* (Berkeley: University of California Press, 1996), p. 364.

41 Robert Smithson in 'Four Conversations between Dennis Wheeler and Robert Smithson (1969–70)', Flam, *Robert Smithson*.

> get there you won't really know where you are. In a sense the non-site is the centre of the system, and the site itself is the fringe or the edge.[42]

And:

> Right, because also the site isn't there. Here you are confronted with both a mental and a physical manifestation that purports not to be there, so that's an effacement through the physical properties of both mind and matter. In other words, in the non-site idea here is this container and it has the limit of my mental experience plus the physical point.[43]

Spiral Jetty is a complex system of sites (and non-sites) which offer themselves up for *observation* as a system of distributed representation. And the ambiguity regarding which of the various aspects of the system to be prioritised becomes the central feature of the work. As Owens observed:

> Like the nonsite the Jetty is not a discreet work, but one link in a chain of signifiers which summon and refer to one another in a dizzying spiral. For where else does the Jetty exist except in the film which Smithson made, the narrative he published, the photographs which accompany that narrative, and the various maps, diagrams, drawings, etc., he made about it? Unintelligible at close range, the spiral form of the Jetty is completely intuitable only from a distance, and that distance is most often achieved by imposing a text between viewer and work… That Smithson thus transformed the visual field into a textual one represents one of the most significant aesthetic 'events' of our decade; and the publication of his collected writings constitutes a challenge to criticism to come to terms with the textual nature of his work, and of postmodernism in general. That challenge is formidable, since it requires the jettisoning of most of our received notions about art; it can only be acknowledged here.[44]

Given this complexity the key questions to ask of this complex system – *Spiral Jetty* – are the questions that were also raised in the preceding discussion concerning representation and the work of Jackson Pollock. Thus the significant question to ask of the work is not

42 Smithson in conversation in 'Discussions with Heizer, Oppenheim, Smithson (1970)', ibid., p. 249.

43 Ibid., p. 204.

44 Craig Owens, 'Earthwords', in *Beyond Recognition* (Berkeley: University of California Press, 1992), p. 47.

only *what is being represented*; but also *how is the work representing*; and also further *how is the work itself observed* as a system of distributed representation from within particular discursive systems?

Spiral Jetty is observed as a complex system in so far as it is a system defined by Ackoff in 1960 as:

> The term system is used to cover a wide range of phenomena. We speak, for example, of philosophical systems, number systems, communication systems, control systems, educational systems, and weapon systems. Some of these are conceptual constructs and others are physical entities. Initially we can define a system broadly as any entity, conceptual or physical, which consists of interdependent parts. Even without further refinement of this definition it is clear that in systems research we are interested only in those systems which can display activity – that is behavioural systems.
>
> The essential characteristic of a behavioural system is that it consists of parts each of which displays behaviour. Whether or not an entity with parts is considered as a system depends on whether or not we are concerned with the behaviour of the parts and their interactions.[45]

Further, *Spiral Jetty* is also complex according to Luhmann's definition of complexity:

> When the number of elements that must be held together in a system or for a system as its environment increases, one very quickly encounters a threshold where it is no longer possible to relate every element to every other one. A definition of complexity follows from this: we will call an interconnected collection of elements 'complex' when, because of immanent constraints in the elements' connective capacity, it is no longer possible at any moment to connect every element with every other element. The concept of 'immanent constraint' refers to the internal complexity of the elements, which is not yet at the systems disposal, yet which makes possible their 'capacity for unity'. In this respect, complexity is a self-conditioning state of affairs; the fact that elements must already be constituted as complex in order to function as a unity for higher levels of system formation limits their connective capacity and thus reproduces complexity as an unavoidable condition on every higher level of system formation.

45 Russell Ackoff, 'Systems, Organizations, and interdisciplinary Research', *General Systems Yearbook*, vol. 5 (Society for General Systems Research, 1960), p. 3.

Leaping ahead, we may hint at the fact that this self-reference of complexity is then 'internalised' as the self-reference of systems.[46]

This idea that it is impossible to relate every element to every other one except by virtue of their interconnectedness suggests that complexity is about the unity of system as a whole which is of greater conceptual significance than the sum of the parts.

In other words a complex system demonstrates behaviour which is unpredictable and which can't be quantified by an analysis of the separate constituent parts of the system as discreet elements. Thus the whole of the system is a separate theoretical entity which cannot be described purely in terms of its interactions of interconnected parts. This is the idea that complex system display emergent properties which arise from out of their structural organisation.

In complex systems representations are distributed over the whole system, and do not reside in a single part of the system. Hence in cognitive models of the mind as connectionist/distributed system demonstrate memory in a connectionist manner; and if part of the brain is destroyed this does not correspond to the loss of a particular memory. Instead, when damaged, due to their distributed nature, distributed systems display what is known as 'gentle degradation' of their representations. Following from this it is argued that despite its physical degradation the complex system of *Spiral Jetty* can still be observed via the systems across which its representation is distributed; and this is despite the 'gentle degradation' of its sculptural form.

In conclusion; the physical volatility of the sculptural aspect of *Spiral Jetty* (illus. 12) reminds us that it is not a single thing. It is, instead, a complex system from which *Spiral Jetty* becomes observable as an emergent property.

46 Niklas Luhmann, *Social Systems*, trans. John Bednarz and Dirk Baecker (Stanford: Stanford University Press, 1995), p. 24.

Chapter 5: The Gallery as System

Introduction

In this chapter I identify that the systems of display and preservation of art are complex systems. In doing so I apply two key systems theoretical concepts to the systems of the art gallery:

1. Complex systems are *distributed* systems of representation. That is that the representations associated with such systems are *distributed* across the entire system. In the case of the art gallery I argue that is demonstrated by the concept of art itself being understood as a function of a representation which is distributed throughout the gallery system.[1] In this respect I argue (as in Chapter 2) that the institutional theory of art can be restated from a systems theoretical perspective.

2. Complex systems display the behaviour of *positive feedback.* That is they become 'locked-in' around particular patterns of behaviour. In the cases of the gallery systems outlined below I argue that when this complex systemic 'locking-in' is manifested in cultural systems it leads to the 'locking-in' of certain cultural tropes. In other words the systems-theoretical principle of positive-feedback and irreversibility leading to a 'lock-in' within complex systems can be used to explain patterns of cultural inertia and structures of social hegemony.

In order to demonstrate more thoroughly the systems theoretical concept of 'lock-in' I have already advanced a discussion of positive-feedback, irreversibility and lock-in as it was used by W. Brian Arthur to explain the development of the VHS/Betamax video market which became 'locked-in' around VHS as the dominant (but not necessarily technologically superior) format. My conclusion is that the complex

1 The 'gallery system' refers to the individual gallery space as well as to the worldwide network of galleries of which individual galleries are a subsystem.

systems of the display, preservation and archiving of art can also be said to display isomorphic, complex 'locking-in'.

Further it is also noted that in the spirit of the book as a whole; namely that isomorphisms can be identified across a wide variety of complex systems – the arguments in this chapter concerning the art gallery/museum as a complex system can also be applied to both the artworld and art history which are also complex systems. And in the chapter immediately following this one I argue that, like an art gallery, art history is a 'museum of language' which is also a complex, distributed representational system which is 'locked-in' around certain patterns of behaviour, cultural tropes and social hegemonies.

1. The 'Genuine Fakes'

What follows in this chapter section is a cautionary tale for revisionist art-historians: be careful not to break the law.

On Friday February 12th 1999 Judge Geoffrey Rivlin convicted John Drewe of conspiracy to defraud, forgery, theft and using a false instrument with intent. The following Monday he was sentenced for 6 years. The severity of his crime was deemed to be so grave that Detective Sergeant Jonathan Searle of the Organised Crime Squad was moved to state that: 'The World will be a much safer place with Drewe in jail. It is a relief to know this man will now face the justice he deserves.'[2]

Drewe's allegedly heinous crime which was identified as such an affront to decent humanity is something which all art historians should be vigilant of. 'Anything he has touched becomes suspect', Detective Sergeant Searle elaborated. 'He has rewritten art history.'[3]

Drewe had apparently made art out of nothing. His rewriting of art history was deemed to be criminal because he had seemingly

2 Amelia Gentleman, 'Fakes leave art world in chaos', *The Guardian* (Feb. 13, 1999), p. 10.

3 Ibid.

created works of art where none should have existed. He was punished because he had tampered with the complex internal workings of the gallery system and its archive. He had, in short, been able to alter what was observed as art by this complex system.

The story began in 1987 with an advertisement placed by the painter John Myatt in the back of *Private Eye*. Myatt advertised his services as offering: 'Genuine fakes. Nineteenth and twentieth century paintings from £150. Phone [...]'[4]

Since 1981 the painter Myatt had earned a humble living by counterfeiting artworks in a variety of modern styles, as he himself explained: 'Quite a few people wanted impressionist paintings. The only thing I avoided as Pre-Raphaelite work, because it was too detailed and took too long.'[5]

Myatt's work was also characterised by his use of cheap materials which are easily available from hardware stores. His 'genuine fakes' were made using Leyland trade paints (emulsion), petroleum jelly and powder paint.

When Drewe contacted Myatt the counterfeit and fraud partnership began. Firstly Drewe asked Myatt to paint a Matisse-style painting. Satisfied with the result he then requested several other styles of modernist painting, leaving Myatt free to choose artist and style.

Later that year Drewe took a counterfeit Gleizes to an auctioneer who identified the Cubist work to be either a Gleizes or after Gleizes and valued the work at £12,500. This set the whole process in action with Myatt producing his genuine emulsion fakes and Drewe circulating them within the systems of the artworld. At the trial the jury heard that by the time the partnership were apprehended, in 1995, Myatt had created by his own estimate more than 180 fakes. It was also estimated that Drewe may have earned more than £2.5 million from his conspiracy.

The problem which Drewe presented to the system of the artworld, and presumably the reason as to why it was alleged that the world is now a 'much safer place' with his subsequent incarceration, is the way in which he gained credibility for his fakes. A question

4 Ibid.

5 Jay Rayner, 'The Fake's Progress', *Observer Review* (July 4th, 1999), p. 7.

mark still hangs over the whereabouts of all the genuine fakes and just how much money was made. As Detective Constable Micky Volpe said: 'The corruption which has taken place of the national archive of contemporary art has been considerable.'[6] And it may be that there are still Drewe and Myatt's simulacral art works are still being circulated around the circuits and systems of the artworld.

Above I argued that the plausibility of Drewe's simulacral art works can be explained by the ways in which the art gallery system represents works of art; that is as a complex system of *distributed* representation. The confidence trick worked because Drewe was able to redirect information and thus provide opportunities for the genuine fakes to be represented over the gallery system as authentic. This was achieved by creating false provenances for the genuine fakes. I will now outline how this 'false' representation was made possible.

As the agent by which the genuine fakes entered into circulation in the artworld system the first thing that Drewe needed to do was to draw attention away from himself. To do this Drewe created false owners like Daniel Stoakes who he coerced into selling paintings for him and who himself faced trial at the time for his involvement. Whilst Stoakes was a real person, hoodwinked into the fraud as a front for false provenaces and sales, another 'owner', Len Martin, never even existed. Using correspondence with a professor of physics at Kiel University, in Germany, called H. Heinrich Martin Drewe reconstructed letters of authentication using Professor Martin's signature. Prosecutor John Bevan QC explained:

> What he did was take an original document from a solicitor's, skillfully cut-and-pasted it, whited out the lines, photocopied it, copied it again. There you are, a lovely official document: 'To Whom It May Concern, this letter confirms the painting by Giacometti is being sold with the full authority of the owner.' This letter is signed by the owner of the Giacometti, Len Martin. Len Martin doesn't exist.[7]

Drewe then approached the Institute of Contemporary Arts in London and donated two forgeries, a Le Corbusier and the Giacometti,

6 Gentleman, 'Fakes leave art world in chaos', p. 10.

7 Peter Landesman, 'A 20th Century Master Scam', *New York Times Magazine* (July 18th 1999).

to be sold at an auction to raise funds. Given the apparent generosity of his gifts the scrutiny directed toward both himself and the art works was less strict. The gifts meant that Drewe was able to pose as a patron with a particular interest in the institution; he was thus able to gain access to the archives at the I.C.A. and he was given a key to the library. Drewe made a similar offer to the Tate gallery and offered it a pair of forged paintings apparently by Bissire. Again there seemed to be an equal desire for accepting the works as a genuine act of generosity. However when the paintings were shown to Bissire's son he threw the works' authenticity into doubt by questioning the materials used. In order to avoid a potentially difficult situation as the Tate stalled on accepting the work Drewe withdrew the paintings in favour of a £20,000 cash donation. Again Drewe used the gift in order to pose as someone with a serious interest in the archive (the Tate funds were offered to support the upkeep of the catalogue,) and again Drewe used the gifts as leverage to apply for access to the archives to which the Tate readily agreed. Drewe demonstrated similar duplicity to gain access to a third archive – the Victoria and Albert's National Art Library. In his application for access Drewe listed himself as 'professor of physics', an identity which he had travelled under for some time. He listed his own address as the contact for a character reference for 'Dr John Cockett', a fictional character of his own creation; and when The National Art Library requested a reference Drewe wrote his own – 'John Drewe is a man of integrity', he wrote and signed it 'John Crockett.'

Drewe used his access to the archives to aid his fraud. In the police reports of the case it was found that Drewe had used his access to the I.C.A. archives, where he was given a key to the library, to create false provenances for the works and paperwork bearing the I.C.A. letterhead. Correspondence from their archives featured in some of the false evidence which Drewe used to support the sale of the genuine fakes. In the National Art Library at the Victoria and Albert Museum Drewe also doctored the archives. One example involved the counterfeit Giacommetti first offered to the Tate which he validated by altering an old catalogue. Taking a catalogue from 1955 of an exhibition called 'Exhibition of Paintings, Sculpture and Stage Designs by Members of the Entertainment World', at the Ohana

Gallery in London, Drewe reprinted several of the pages of the publication to make it appear that his own genuine fakes had been on display at the exhibition (including with the fake Giacometti and some counterfeit works supposedly by Chagall, Dubuffet, de Stael and Nicholson.) He then placed his own pages (which he had printed on a computer and aged accordingly) into the catalogue and re-stitched the catalogue binding thus creating a permanent archival record of the existence of the genuine fakes which could be used to validate sales. A similar level of tampering is believed to have taken place at the Tate; and when detectives first raided Drewe's house on April 6th 1996 they found several rubber stamps of authentication bearing the Tate seal.

Drewe's fake documentation proved to be highly plausible even when used to support a lacklustre artwork. As Peter Landesman observed:

> Alan Bowness, former head of the Tate and the son-in-law of Ben Nicholson, was fooled into authenticating two of Myatt's fake Nicholsons, not because the pictures were good – in fact, the general consensus was that they were unimpressive at best – but because the provenancing was flawless.[8]

In addition to his contamination of documented evidence for the existence of the forgeries Drewe also set up his own validation agency called Art Research Associates which offered a professional provenance checking service. One New York dealer had obtained the 'genuine-fake' Giacometti (for $175,000) but was having trouble selling the painting on as the Alberto et Annette Giacometti Association under the directorship of Lisa Palmer refused to vouch for the work. Via a middle man (an accomplice of Drewe's who posed as a representative of the Art Research Associates) the dealer was put in touch with Drewe. Posing as an expert and calling himself Professor Drewe, Drewe was then able to show the dealer the faked Ohana gallery catalogue in the National Art Library which included black and white photographs which proved the paintings was genuine. Further evidence included faked letters from owners who had never existed and many counterfeit bills of sale which charted a virtual history of the

8 Ibid.

work throughout its circulation within the system of the artworld. In support the Art Research Associates produced a report in which it was claimed that the painting had been 'thoroughly investigated' some months previously and that subsequently, 'the provenance of this painting is not in question'. The dealer was subsequently convinced.

It is therefore argued that galleries and their archives are complex representational systems and that Drewe had corrupted these gallery systems with the corrupted or 'dirty' data which mis-attributed works of art. In doing so, I argue, he was able to exert an influence over the representative function of these complex systems. In short he was able to make these systems represent his own fakes as genuine.

It has been estimated that of the approximately 200 works painted by Myatt and circulated by Drewe the whereabouts of only 73 are known. The rest are still in circulation in the art world; being bought, sold, gazed upon and consumed.

It is further observed that the phenomenon of the 'genuine fake' is not particular to this particular case. It has been claimed that between 10 and 40 percent of pictures by major artists are faked or have been restored to such an extent that they are what can be deemed the equivalent of a fake.[9] Landesman observed that:

> One of a number of reputable West End dealers drawn into [the Drewe and Myatt] scam, who demanded anonymity, claims that 15 percent of sales at auction houses are fake. Thomas Hoving, former director of the Metropolitan Museum of Art, has said that during his tenure, a full 40 percent of the artworks considered for purchase by the museum were phony or overrestored. Geraldine Norman, the British art journalist credited with exposing Keating and Hebborn, says that at least 10 percent of the pictures purportedly made by major Impressionists are fake. Some experts say that as much as 60 percent of the Giacomettis on the market are fake, at least in part as a result of Drewe's con. Forty percent of pictures supposedly from the Russian avant-garde era are almost certainly forgeries. The widows of Modigliani and Chagall were accused of selling certificates of authentication. Salvador Dali, lying on his deathbed in a stupor, is said to have been fed thousands of sheets of blank paper to sign for

9 Ibid. It might also be the case that the easy acceptance of the genuine fakes was made increasingly possible by the climate of credibility created by the boom in the art market of the 1980s. Given the value of art as a commodity people were all too eager to buy and sell any art, no matter what, as long as someone wanted to buy and someone wanted to sell.

> fake lithographs. In a grotesquely sad twist, the Italian painter Giorgio de Chirico was caught back-dating new pictures to a time, 20 years previously, when his work was popular and in heavy demand, in effect forging himself.[10]

In conclusion of this section I argue that the claim that Myatt/ Drewe's 'genuine fakes' are artworks which are fake yet which pose as genuine is an oversimplification of the complex representation which has taken place. This is because it leaves the problem of identifying what distinguishes the genuine from the fake as unanswered except in rigidly material terms alone (and that such materialism on its own is an unsatisfactory art historical tool not least because it fails to account for more recent artistic practice such as conceptual art and the 'dematerialised' art object of the second half of the 20th century.) What is instead suggested is that the undiscovered genuine fakes are represented by the complex systems of the artworld in exactly the same way as other genuine works. In other words the 'genuine fakes' are not simulacral artworks masquerading as the real thing; instead they *are* the real thing and this can be demonstrated by looking at the ways in which they are being bought, sold, gazed upon and consumed accordingly. This is a further articulation of a key theme of the second half of the book (regarding the application the systems theoretical approach to the arguments of the institutional theory of art and using it to provide a plausible ontology of art in terms of as a representation within the various and interconnected complex systems of the art world.) It is argued that art is not dependent upon a category of material existence but rather is relative to an act of representation or observation; and hence relative to the system of representation (or observation.) I discuss this in more detail below.

10 Ibid.

2. Locked-in: The Art Gallery and Positive Feedback

Having identified that the Drewe/Myatt case demonstrates how the art gallery is a system of distributed representation it is also stated that this brings it within the sphere of arguments common to institutional theories of art. In the arguments which follow I will discuss how the representational system of the art gallery may be identified in terms specific to dynamical systems theory; that is as a *complex* system which therefore displays the behavioural properties of complexity. This is demonstrated through identification of two types of behaviour specific to *complex systems*. These are:

1. Irreversibility and Positive Feedback (or 'lock-in');
2. Distributed Representation.

The Drewe Myatt case has been used to illustrate these two points in the following terms. The ambiguity of the ontology of the 'genuine fakes' highlights the following:

First, the distributed nature of the representations of the art world and in particular the gallery system. Thus the representation of the 'genuine fakes' as genuine is distributed over the gallery system rather than being specifically related to the material existence of the work of art itself.

Second, the role which irreversibility and positive feedback play in the gallery system is highlighted. This is because once an artwork has been 'represented' as a genuine artwork then it is very difficult to prove otherwise and to reverse the process; hence the missing approximately 120 'genuine fakes'.

2.1 Lock-in and Saatchi

The systems of the artworld are, as I argue here, complex systems and thus display behaviour common to complex systems such as being auto-catalytic (that is self-perpetuating,). They do this by being financially and socially self-determining and self-justified. This is demonstrated below by the activities of Charles Saatchi and his role within the contemporary British artworld system of galleries, dealerships and criticism.

Since Charles Saatchi established the Saatchi Gallery in 1985 as a showcase for his collection, the gallery has become a focus for contemporary British Art within the artworld system. As a dealer and collector he has been identified as having considerable influence within the art market. As Dan Glaister noted, one way of looking at Charles Saatchi's influence is as:

> The driving force behind the rebirth of British art, the patron who has provided the locus and the critical mass to turn a disparate group of artists into a movement.[11]

By means of an example on the 8th of December 1998 Saatchi held a sale of a small part of his collection (between 4 and 5 percent). The 130 works by 97 artists secured £1.6 million, with 109 works being sold. The sale created an enormous amount of publicity and speculation in the press over Saatchi's influence within the artworld.

One of the purposes of the sale was to raise money to fund the establishment of bursaries to finance the production of future young, contemporary art;[12] and it has been suggested that it provided a means for Saatchi to use his influence as a dealer and collector to cherry-pick the work of new artists at source which were attractive, collectable works with a high resale potential. As Patricia Bickers noted, in *Art Monthly*, this enabled Saatchi to have:

> The pick of the work of the new London graduates of the future. In this way, Saatchi could be said to have taken control of both the means of production and distribution, which should not surprise us about a collector who has been described… as the Rupert Murdoch of the artworld.[13]

11 Dan Glaister, 'Saviour or Destroyer? Maybe Both', *The Guardian* (Oct. 21, 1998), p. 5.

12 £10,000 a year was offered to Chelsea, Goldsmiths, the Royal College and Slade Art schools in London; the Young Artists' Sponsorship Bursaries were established to provide support for artistic projects.

13 Patricia Bickers, 'Editorial', *Art Monthly*, 222 (Dec. 1998–Jan. 1999), p. 18.

The sale also demonstrated that this was a system which was autocatalytic and self-perpetuating in so far as it could fund itself. By buying art cheap, in bulk, and selling it on for profit Saatchi could use the money to buy future art whilst keeping the size of his collection in check. Saatchi himself noted a desire to make this system self-sufficient and he remarked that: 'the collection has to be self-financing. Otherwise it would be a bottomless pit.'[14] (And it will be observed below, that this system is also self-sufficient in the way it provides the conditions for its own critical and social legitimacy.)

At the time of the Christie's sale the system of the art market was in a particular situation defined by the two characteristics given below.

First, the art market boom of the 1980s was still a recent memory. During this time, aided by the voraciously free-market environment of Thatcherism, Saatchi's personal stock as both an advertising consultant (as the joint chairman of Saatchi and Saatchi) and a speculator on the art market had risen dramatically.

Second, despite a recession[15] the art market continued to perform consistently well as Andrew Marr explained:

> During the last recession, art auction houses enjoyed a two year commercial boom. At the top end of the market, only a couple of weeks after Black Monday with stock markets plunging Van Gogh's Portrait of Dr Gachet was sold by Christie's in New York for \$82.5m.[16]

A major reason for this, as Marr explained, is that art was seen by speculators to be a less risky investment option than stocks and shares:

14 Dalya Alberge, 'A Very Private Collector: Charles Saatchi, Britain's Leading Modern Art Collector Talks (but only a little) to Dalya Alberge', *The Independent* (March 3rd, 1992), p. 19.

15 A world-wide recession followed 'Black Monday' – the notorious October 19, 1987 when the Dow Jones Index fell by 508 points (almost 22%), which was identified as the largest one-day drop in the history of the New York Stock Exchange.

16 Andrew Marr, 'Throwing the Baby Out With the Bath Water?' *The Guardian Saturday Review* (Sat. Oct. 24th, 1998), p. 3.

While I was [at a sale by Jeffrey Archer of 111 works by Warhol and 71 by Hockney at the Peter Gwyther Gallery, off Bond Street in October 1998] a buyer phoned in to say he was taking £50,000 out of shares: he'd rather have paintings. At [another sale of contemporary art at] Sotheby's the place is thronging with South Americans, Italians and Japanese who also feel that, for the time being, art is safer than equities… So art-works become units of value like coffee options.[17]

This is significant from a systems theoretical perspective because, as outlined in W. Brian Arthur's example of the video-recorder market of the 1980's, economies can be seen to behave like complex systems. This means that apart from Saatchi's influence the current art market adheres to certain examples of positive feedback as Marr observed:

There is a network of accepted values and conventions underpinning a painting or print as a tradeable commodity. For a painting, that includes the artist's status in a league table (Picasso equals the US; Pissarro, Germany; and Marquet, Argentina); the quantity of his work in circulation (art monetarism), the existence of similar works in public collections (the national reserve banks.)[18]

Saatchi clearly has significant influence within the markets of the art world. As Gordon Burn observed with regards to Saatchi's influence:

Saatchi's nose for the coming young art star, the next new thing, is legendary by now. He has an unrivalled track record. For a decade his purchasing decisions have influenced collectors of vanguard art all over Europe and America. Being taken up by Saatchi has become one of the conditions of success for any young artist.[19]

A further example of Saatchi's influence is demonstrated by the exhibition *Sensation: Young British Artists from the Saatchi Collection* (1997) which was held at the Royal Academy of Arts in London. With over 300,000 visitors *Sensation* was the most popular RA show

17 Ibid.
18 Ibid.
19 Gordon Burn, 'I Want It, I Want It All, and I Want It Now', *The Guardian G2* (Dec. 7th, 1998), pp. 1–2.

of contemporary art in living memory[20] and it greatly increased the public profile and popular awareness of Saatchi's collection and his activities as a dealer and collector.

Commentators have complained[21] that *Sensation* provided a lop-sided view of the state of British art being based, as it was, on the concerns (financial and aesthetic), taste and whims of just one collector and dealer. From a systems perspective *Sensation* provides an example of a cultural system becoming 'locked-in' around particular pattern of cultural activity and particular objects. In this case the 'attractor' around which the system had become locked was Saatchi himself.

And as Stallabrass observed this influence and 'lock-in' can also lead to a conflation between perceptions of economic and aesthetic value:

> Art objects are near-perfect commodities: lacking the complications of usefulness, they are pure tokens of exchange, their value based on opinion alone. Yet the power of their exchange value is taken for an aesthetic charge... Money is the necessary and often the sole raison d'être of the gallery system; yet, at least in terms of the presentation, the galleries have a prissiness about it, carrying the conventions of mercantile display to such a rarefied level that the visible signs of commerce (price tags, special offers) tend to disappear. Money must simultaneously be gestured at and concealed, since it is both essential to aesthetic aura and what threatens to taint it.[22]

Stallabrass's conclusion is that, what he refers to as the current 'fine art system', there can be no distinction made between aesthetic and economic value; the expensive is regarded as the beautiful, and that deemed to be beautiful will certainly be expensive:

> In pursuing aesthetic value, we embark on the old quest for the ghost in the machine of the fine art system. As that system is currently set up, money is both the fuel of the machine and its output, aesthetic value its exhaust, constantly

20 'Sensation' was 'the Royal Academy's most popular show since a summer exhibition in the dying days of Queen Victoria.' Marr, 'Throwing the Baby Out', p. 3.

21 A thorough bibliography of the literature on Saatchi is provided by Julian Stallabrass, *High Art Lite* (New York: Verso, 1999).

22 Ibid., p. 80.

> evaporating in the polluted atmosphere of mass culture, and as long as the machine runs, constantly replaced. It is pointless to look for the ghost in the machine; look rather for the pecuniary machinc in what appears to be the ghost.[23]

Given this conflation between the aesthetic and the economic those (such as Saatchi) who exert a significant economic influence within this system can therefore also be regarded as exerting a significant aesthetic influence within this system. In this respect the artworld system behaves like a complex system by demonstrating the effects of positive feedback. Further, given his role as a cultural attractor within the gallery systems within which he operates, Saatchi's influence should be seen as not only unsurprising but actually as inevitable consequence of complex behaviour within the art world systems as they become 'locked-in' around particular cultural tropes.

Stallbrass rejected the gallery as the sole provider of art status to objects as too simplistic and also emphasises the role which art schools play in the establishment and definition of Art. He argued that:

> By far the most successful definition – though still imperfect – is to say that art is something done by those who went to art school. The closed shop operates with remarkable effectiveness, and you will find very, very few artists endorsed by the gallery system, private or public, who have not been through an accredited course. The art schools are at the opposite end of the process to the public galleries but both have the same role, that of sanctioning a particular set of objects or practices as art.[24]

This provides a further example of the artworld system locking-in around a particular pattern of behaviour and producing a particular output. In the case of the work of the YBAs supported by Saatchi the role of art schools, particularly Goldsmiths, provides us with clear examples of positive feedback. In this case the art market became locked-in around particular style or ethos which even though diverse and hard to define exactly could be broadly defined in terms of Saatchi's patronage. This is the style which Stallabrass has dubbed 'High Art Lite.'

23 Ibid., p. 83.
24 Ibid., p. 180.

3. Five Implications of the Application of Dynamical Systems Theory to an Analysis of the Cultural Systems of Art

Complex systems have a memory which is expressed in terms of their irreversibility (as discussed in Chapter 2 in reference to how the market 'remembered' initial effects causing it to 'prefer' VHS over Betamax). Complex systems reach bifurcation points or points of no-return beyond which when passed the system cannot be reversed to a previous state. Because of the irreversibility of a system's memory it 'remembers' previous states and incorporates these into its current state. In turn this means that effects cannot be reversed, small causes can be amplified to have big effects; often with surprising and unexpected results such as the dominance of VHS. This demonstrates how a system's memory can lead to the effects of positive feedback or what has also been called the 'butterfly effect'. W. Brian Arthur calls the effects of positive feedback examples of systems 'locking-in' around a certain behavioural pattern. And, I have already noted above, systems such as ant colonies or financial markets can become 'locked-in'.

I will argue, therefore, that the systems of art galleries and art markets can, like ant colonies and video-recorder markets, also become 'locked-in' around certain patterns systemic behaviour and attractors. And via the mechanisms of positive feedback cultural systems, just like other systems, can become more and more firmly entrenched in representing particular values and interests and thus become *locked-in*. This leads me onto the concluding section of this chapter in which I outline 5 implications of the application of dynamical systems theory to an analysis of the art gallery. They are presented below.

Saatchi is just one example of a strange attractor around which a complex social system has become locked via the processes of positive feedback. Von Bertalanffy identified a key aspect of his General Systems Theory as a similarity ('isomorphism') to be found between disparate systems. If the example of Saatchi's influence is a demonstration of the behaviour of one complex system then, given these suggested isomorphisms between systems, we should be able to find other examples of cultural 'locking-in' in gallery systems. As the example of Saatchi demonstrated a gallery system can be seen to lock-in around the dominant personality of a key collector. The importance of key collectors has shaped the accepted histories of art as Gere and Vaizey observed:

> Collecting must, in our view, significantly alter the repute of the objects collected, not only by adding to knowledge and expanding appreciation, but perhaps even more by conferring status: the collector can make the unfashionable or ignored more central to the culture of the day.[25]

To demonstrate this, a similar process of feedback and 'lock-in' can be seen in the significance of Peggy Guggenheim's collecting in the establishment of American Abstract Expressionism as a dominant cultural system. Like Saatchi she was a wealthy patron of contemporary art whose support, especially whilst she lived in New York and collected American Art, could transform the status, significance and value of an artist. This is demonstrable in the example of Pollock whose reputation was, arguably, created by the dual poles of influence of Peggy Guggenheim and Clement Greenberg. Tom Wolfe described this positive feedback loop of how an artist's reputation could snowball on receiving the correct patronage in the following terms:

> Greenberg didn't discover Pollock or even create his reputation, as was said so often later on. Damnable Uptown did that. *Pick Me!* Peggy Guggenheim picked Pollock. He was a nameless down-and-out boho Cubist. She was the niece of Solomon (Guggenheim Museum) Guggenheim and the centre of the most chic

25 Charlotte Gere and Marina Vaizey, *Great Women Collectors* (London: Harry N. Abrams, 1999).

Uptown art circle in New York in the 1940s… In a single year, 1943, Peggy Guggenheim met Pollock through Baziotes and Motherwell, gave him a monthly stipend, got him moving in the direction of Surrealist 'automatic writing' (she loved Surrealism), set him up on Fifty-seventh street – Uptown Street of Dreams! – with his first show – in the most chic Modernist salon in the history of New York, her own Art of This Century Gallery, with its marvellous Surrealist Room, where the pictures were mounted on baseball bats – got Sweeney to write the catalogue introduction, in prose that ranged from merely rosy to deep purple dreams – and Barr inducted one of the Paintings, *The She Wolf*, into the Museum of Modern Art's Permanent Collection – and Motherwell wrote a rave for Partisan Review – and Greenberg wrote a super-rave for *the Nation*… and, well, Greenberg was rather late in getting into the loop, if anything.[26]

3.2 The effect of individual agency is minimised

In identifying the artworld and the gallery as systems they are thus identified as complex interconnection of component parts. An ant colony, for example, might be seen to behave like such a complex system of many connected components. In the ant colony, as with other complex systems, the organic unity of the whole system is of greater operative significance than the individual components. Thus it is not that one smart ant in the maze finds the way to and from the food, similarly in the complex system of the brain one neuron doesn't think for itself but instead thought is distributed over the entire system. Likewise in W. Brian Arthur's example in order to explain the market shift to VHS we can't explain this through the actions of any singular agent, be it the VHS designers, their marketing department or a particular recorder dealership. Instead we can note that the system as a whole settled into a particular pattern. Similarly whilst it may be rhetorically convenient to lay the blame for the current state of contemporary British art solely upon the shoulders of one man this is as unsatisfactory as blaming an art for the behaviour of its colony, or blaming a neurone for an inappropriate thought. Within any complex system and hence also within the artworld the effect of individual agency is minimised. Hence we should look upon the phenomenon of

26 Tom Wolfe, *The Painted Word* (London: Black Swan, 1989), p. 56.

Saatchi as distinct from Saatchi the individual. As a phenomena it is one which is relative to the systems within which emerges; in other words as an attractor rather than an agent; and an emergent phenomena rather than prime mover. And within such complex systems there will be other attractors and emergent phenomena. As an individual however we should see Saatchi the man and other figures and components like him, out of which the whole network is composed, as just singular elements which contribute to the functioning of the greater unity of the system. Just like an neurone or ant.

3.3 Criticism becomes self-fulfilling

Every damning column inch, book or radio and television report which condemns Saatchi (as in Hatton & Walker's *Supercollector*),[27] his taste in British Art (Stallabrass) and his influence contributes to a self-fulfilling cycle of positive feedback. Hence with every act of reflection the system becomes further 'locked-in' around these examples. News breeds news. And the media phenomenon surrounding Saatchi features prominently in most of the commentaries upon him. This, in turn, creates more column inches and publicity which creates more of a story; and so on, and so forth.

The Royal Academy was complicit in the generation of positive feedback. It courted the press with information about the controversial works within the show such as the exhibition's show-stopper Marcus Harvey's *Myra* (1995) which received fierce criticism from the right-wing press over its portrayal of Myra Hindley painted using a cast of a child's hand; Chris Ofili's *The Holy Virgin Mary* (1996) also received strong criticism.[28] The Academy also publicly suggested that some of

27 Rita Hatton and John Walker, *Supercollector* (London: Ellipsis, 2000) which is a crude attempt to provide a self-styled 'hostile critique written from a socialist perspective', p. 10.

28 See 'Mother's Fury at Myra "Art"', *Daily Mail* (July 26th, 1997); Jeremy Armstrong and Paul Byrne, 'Artless: Fury at Child-Fingerprint Portrait of Monster Myra', *Mirror* (July 26th, 1997); Alexander Waugh, 'The Great Painting of Myra "Designed to Shock"', *Evening Standard* (July 25th, 1997); Bill Mouland, 'Invitation to Outrage', *Daily Mail* (Sept. 17th, 1997); 'Exhibited by the

the artists within the show were to be offered RA membership.[29] This offer was a hollow and cynical gesture calculated to generate further publicity[30] because, as Stallabrass noted; firstly the number of Academy members is fixed so that normally members will only be replaced on death. And secondly the offer was to provoke an inevitable yet unwitting complicity in the positive feedback loop of publicity from the artists themselves who immediately and predictably condemned the RA board; Hirst called them 'Fat, stuffy, pompous' and Jake Chapman branded them, 'whinging alcoholic old farts[s]'.[31]

The self-perpetuating tension in the RA board was replicated by the oppositional opinions exhibited in the critical responses to the exhibition with a conflict between those seeking to condemn and the exhibition's apologists. However, the exhibition was not ignored, and both negative and positive responses served the same function of ever increasing interest with critics either endorsing the controversy or creating more of the activity they were seeking to oppose. As Stallabrass said, the work actually requires opposition in order to create a significance for itself: '*High Art Lite* has relied upon the outrage of such red-faced bored to make it seem young, cool and oppositional.'[32]

As Gordon Burn observed, *Sensation* earned a great deal of publicity for how much publicity it was gaining.[33] This is a technique with which Saatchi would have been familiar given his experience as an advertising executive and it has echoes of the advertising strategy of embedded marketing whereby a product is planted in the media in the

Royal Academy in the So-called Name of Art, Defaced by the People in the Name of Common Decency', *Mirror* (Sept. 19th, 1997); 'How can this pile of Sh*t be worth £250,000?', *Sport* (Sept. 19th, 1997); Paul Johnson, 'An Obscene Picture and the Question: Will Decency or Decadence Triumph in British Life?' *Daily Mail* (Sept. 20th, 1997). Stallabrass also provides a good overview of the outrage generated by *Sensation* in 'Saatchi and Sensation' (ch. 7), *High Art Lite*.

29 Stallabrass, *High Art Lite*, pp. 200–202.

30 Although four members did resign over the exhibition.

31 Both references from Stallabrass, *High Art Lite*, p. 202.

32 Ibid p. 209.

33 Gordon Burn, 'I Want It, I Want It All, and I Want It Now', *The Guardian G2* (Dec. 7th, 1998), pp. 1–2.

trojan horse of another story,[34] hence providing publicity without the need to rely upon the placing of expensive adverts. This use of publicity to generate publicity is another form of positive feedback at work and in the case of *Sensation* the brand being promoted was Saatchi himself around which a cultural system was becoming 'locked-in'. Thus commentaries, critical or otherwise, on Saatchi's activities merely perpetuate the increasing phase-lock within the system. In this respect there is a certain irony in observing that the more vociferous a critique the more it will become self-defeating by re-enforcing the cultural significance and positive feedback loops of significance and exclusion which they seek to criticise.

As physical objects books have a permanence and irreversibility within social, cultural and epistemological systems in so far as they both create and curate historical legacies for the material they contain. What is meant by this is that if a book about a particular subject is published and then bought by a library a certain 'locked-in' permanence is granted the subject matter. This is due to the relative irreversibility and permanence of the physical object of the codex. If we take the example of Charles Saatchi and *High Art Lite*, when the current debates concerning the status and role of Saatchi in the contemporary art market have abated, then the book will remain on the shelf for future reference radiating a transfigurative and emergent historical significance onto its subject matter; hence there will be a 'lock-in' in the cultural system around this paradigm. Therefore art without a permanent textual record of its existence (other than itself) around which for the system to 'lock-in' seems much more likely to fade from public and historical memory. In the case of British art of the last twenty years it seems likely that this forgotten art, the locked out Betamaxes of the cultural system will include much of the art ignored or dismissed by Saatchi as being either a poor investment or simply not fashionable or 'cool' enough.

Contemporary artists are well aware of the significance of Saatchi's influence. One of the reasons for this can be attributed to the media-savvy which is indoctrinated into them at art schools such as Goldsmiths where students are instructed into how to succeed in the

34 Such as the example of a scientific study carried out on behalf of a product.

contemporary market. Goldsmiths has produced many of the artists Saatchi has collected in bulk and at a speech made at the Turner Prize award ceremony in the Tate in 1994 Saatchi observed:

> Now I don't know, anymore than you, what has made our young artists suddenly so supercharged, but I am certain that teachers like Michael Craig Martin at Goldsmiths College have had a tremendous effect. Now Goldsmiths, and in recent years the Royal College, and some other art schools, they have set a tone and a level of ambition that has stimulated not just their graduates, but young artists all over Britain who are producing work that is challenging, articulate and relevant.[35]

From this it is not difficult to make the assumption that ambitious students, aware of the significance which Saatchi's influence holds, could be creating works in attempts to appeal to his taste. Chris Offili put it this way:

> A lot of artists are producing what is known as Saatchi art... You know it's Saatchi art because it's one off shockers. Something designed to attract his attention. And these artists are getting cynical. Some of them with works already in his collection produce half-hearted crap knowing he'll take it off their hands. And he does.[36]

Stallabrass made the acute observation that if there is a style to the work of the YBAS it lies in their attempt to produce work that will appeal to the taste of a man whose background in advertising is manifested by a taste which is: 'very much for art that looks like advertisements, and who – except an adman – would want to own one of them?'[37] Thus there is a lock-in around an art which is striking – in so far as it must have the initial impact of a billboard (such as the work of Hirst for example) and work produced in non-threatening and non-volatile media which must be portable, durable and hence sellable. This discreteness of media need not be the only concern, as was seen in the case of Hirst's *Away from the Flock* (1994); a dead lamb supported in a case or Marc Quinn's *Self* (1991); a cast of the artist's head made from nine pints of the artist's frozen blood and a refriger-

35 *Without Walls*, Channel 4 Television (22nd Nov., 1994).

36 Quoted in Stallabrass, *High Art Lite*, p. 200.

37 Ibid., p. 201.

ation unit which must be kept turned on at all times so as not to defrost. However in both cases the relative expense and volatility of the objects extenuated the predominant aspect of the work which provides a unifying aspect to most, if not all of his collection. This is that the work is object based art which provides a positive and easily readable advert for itself.

3.4 Unpredictability

Whilst we may be able to determine patterns in the ways in which apparently chaotic and complex systems behave (such as Lorenz Attractors) they are defined as being unpredictable; we can't know in what direction they will develop. Thus, as W. Brian Arthur noted, despite there being good reasons to predict its success, in terms of a suggested technical superiority over VHS, the Betamax video format was eclipsed in its market. This is because due to the operation of positive feedback within complex systems what appear to be small causes may have very large effects; and we can't know what these effects will be. This leads to two related observations.

First that it may be seemingly insignificant reasons was to why one work of art gets sold over another. This contradicts any belief in any historical teleology within the development of art or that the history of art may be shaped by the desire and prerogative of forceful and dynamic individuals.

Second the issue of quality in art becomes more ambiguous. That VHS would triumph over Betamax was impossible to predict and the eventual outcome was not an index of the technical triumph of the VHS format. Likewise it might also be suggested that particular cultural forms (be these the work of an artist or a particular style etc.) do not succeed, in terms of their social and cultural dominance, because of any supremacy of their technical or aesthetic quality. The Art we have now is merely a product of that around which its cultural support systems have become 'locked-in'. And this leads me on to the fifth, observation in this chapter.

As Peter Davis astutely parodies in his work *Hot One Hundred* (1996) (illus. 13) the canon of what is fashionable is an arbitrary list of the usual suspects which reflect whims of taste which may behave, like other systems, chaotically and complexly. The history of cultural criticism is composed of attempts to redress the balance of canons which have become 'locked-in' around particular cultural forms at the expense of the exclusion of other cultural forms. This suggests that there is a flipside to the institutional theory of art. This complementary view argues that just as there may be a social ontology of art explicable through inclusion within the distributed representation system of the artworld, so too there may be art which has been rejected and excluded from this system (by virtue of not being observed by that system.) To stretch the video-recorder analogy further, if the YBAs and Saatchi-supported art is VHS, then there may also be Betamax art and artists.

It is difficult to find extensive examples of artistic activity which has been completely excluded from the artworld because it is, by this very exclusion difficult, if not by definition impossible, to find. The history of art is a patchwork whose very warp and weft is woven from the strands of that which it has excluded, ignored and rendered obsolete in one form or another. Various politically motivated attempts to redress this balance have formed the basis of many different strategies which have sought to reconfigure or critique the values of the artworld on the basis of gender, class, race and sexuality. The art of the 'outsider' or 'other' is where we will find our Betamax artists.

One example of that which we might call a Betamax artist has been publicised by the artist in question himself who has complained vociferously and at length about his falling stock in both cultural and fiscal capital. The artist in question, Sandro Chia, could have been forgiven for thinking that he would corner a market share when, in 1985, Saatchi bought six of his works. He was, however, sceptical of the attention it would bring and wrote in a letter to *Art News*:

> Mr Saatchi started to buy my paintings very early – so early that I think I was the first painter that was not American to be included in his collection. At the

time of his acquisitions, I remember him telling me how highly important he considered my work, etc. Far from being impressed by his opinion, I rather expressed to my dealers my concern about any person accumulating my works in such number, thereby establishing a control and quasi-monopoly.[38]

When Saatchi subsequently sold all of his works by Chia (illus. 14) his work rapidly devalued as it became, effectively, excluded from the canon. At least this was the way in which Chia saw his status mutate from a VHS to a Betamax artist almost overnight. Writing about the effect of the removal of his work from Saatchi's holding Chia was to note at a later date that:

> At that moment I understood the power of the guy. I thought I might as well go and paint in Alaska. My telephone was ringing constantly and from that moment no one was ringing any more, it was unbelievable.[39]

4. 'Lock-in' as Cultural Inertia – Further Issues

In this concluding section I discuss further the systems theoretical concept of 'lock-in' (or positive feedback and irreversibility) and how it is manifested as inertia of the artworld system.

4.1 The Gallery Building as System: Beaubourg and Baudrillard

The galleries and museums of the artworld are housed in buildings; and the architectural fabric of these buildings will also evolve and adapt like a complex physical system. In *How Buildings Learn* Stewart Brand described how evolutionary architectural systems are at play in the buildings which surround us, and how these buildings not only

38 Sandro Chia, 'Saatchi's "Dispersals"', Letter to *Art News*, vol. 84, no. 7 (Sept. 1985), p. 9.

39 Chia, Statement in 'Making Art, Making Money: 13 Artists Comment', *Art in America*, vol. 78, no. 7 (July 1990), p. 138–139.

– *lock-in* to certain patterns but also form patterns which evolve over time. He says:

> Between the dazzle of a new building and its eventual corpse, when it is either demolished or petrified for posterity as a museum, are the lost years – the unappreciated, undocumented, awkward-seeming time when it was alive to evolution. If [Brian] Eno is right [that buildings demonstrate an 'internal complexity that show the traces of an interesting evolution'], those are the best years, the time when the building can engage us at our own level of complexity.[40]

Brand's argument is that buildings are systems which will change according to their needs and function regardless of the desires of the architect. For this reason he argues that more flexibility should be built into architectural systems from their genesis in order that buildings can improve over time rather than become dysfunctional and obsolete. If a building system cannot adapt well to the change it will experience over time, that is, if it is too 'locked-in' to allow for evolutionary growth, then it will fail; and this is usually manifested in the building being demolished. The alternative, Brand argued, is an architectural system which incorporates change over time, that is the building functions as an adaptive complex system, rather than a rigid, unchanging structure.

Brand noted:

> An adaptive building has to allow slippage between the differently-placed systems of *Site, Structure, Skin, services, Space Plan and Stuff.* Otherwise the slow systems block the flow of the quick ones, and the quick ones tear up the slow ones with their constant change [...].
>
> Age *plus adaptivity* is what makes a building come to be loved. The building learns from its occupants, and they learn from it.'[41]

With its network of exposed services and complex system of pipes, nuts bolts and scaffolding The Pompidou Center at Beaubourg in Paris,

40 Stewart Brand, *How Buildings Learn: What Happens After They're Built* (London: Penguin, 1997), p. 11. The Brian Eno quote is taken from Brand's preceding text and is not attributed.

41 Ibid., pp. 20, 23.

known by locals as 'The Pompidolium',[42] is a gallery which exemplifies Brand's arguments by actually looking like a system (illus. 15). It has, for example, multiple and complex, functions, as Nathalie Heinich documented:

> It is complex in a literal sense in that the building contains several distinct departments, such as the Public Information Library (BPI), the National Museum of Modern Art (MNAM), and the Center for Industrial Creation (CCI), as well as being the home of an number of other activities (children's studio, cinémathèque, meetings/forums, audio-visual unit, publications and periodicals) and having the institute for Acoustic and Musical Research and Creation (IRCAM) close at hand in the *Place Stravinsky*.[43]

The building system not only looks like a machine, it works like one too, illustrating the functioning of the gallery as a processing system. It is a system which processes people and culture in an efficient manner, perhaps too efficient as has been observed:

> Forecasts based on studies carried out in the early 1970's suggested that there *would* be between 8,500 and 15,000 visits per day, making an annual total of between 2.5 and 4.5 million. In reality, the figure has been much higher, the daily average being some 25,000 visitors, giving an annual total of 7.3 million visits.[44]

And

> Since it opened in 1977, 145 million visitors have come to ride the external escalators with their cinematic views of central Paris, and to see great exhibitions of Dali, Matisse and Derain. That's an average of 23,000 a day, and all this for a cultural center designed for an optimistic capacity of 5,000 a day. No wonder the Pompidou was looking shabby and threadbare by the mid-90s.[45]

42 'Not for nothing does local humor call it the Pompidolium – it does in fact look like an oil refinery', Reyner Banham, 'Enigma of the Rue du Renard', *Architectural Review* 963, vol. CLXI (1977).

43 Nathalie Heinich, 'The Pompidou Centre and its public: the limits of a utopian site', Robert Lumley (ed.), *The Museum Time Machine* (New York: Routledge, 1988).

44 Ibid.

45 Jonathan Glancey, *The Guardian* (19 Jan., 2000).

Like other museums, galleries and exhibition spaces *Centre Pompidou* is a factory for culture; an industrial system, and a cultural system which exposes its own supporting structure literally and suggestively – if you know where to look. It is a system with a production line which processes people and generates the appearance [simulacra] of culture, as Baudrillard observed in his extended discussion on Beaubourg:

> Frankly the only contents of Beaubourg are the masses themselves, which the building treats like a convertor, a black box, or in terms of input/output, just like a refinery handling petroleum products or a flow of raw material. Never has it been so clear that the contents – here culture, elsewhere information or merchandise – are merely the ghostly support for the opposition of the medium whose function is still that of beguiling the masses, of producing a homogenous flow of men and minds.[46]

Yet for Baudrillard it is precisely this [dys]-functional efficiency of the institution which provides its potential and its salvation. And this occurs through the institution's paradoxical role in an apparent implosion and negation of cultural activity. In 1977 Baudrillard gleefully (and apparently correctly) predicted The Pompidou Centre's future problems:

> Critical Mass. Implosive Mass. Above 30,000 it threatens to 'Buckle' Beaubourg's structure. That this mass, magnetised by the structure, should become a factor of potential destruction for that very structure… what if this were intended by those who conceived the project (but it is beyond one's hopes)… if it were part of something they had programmed, the chance to finish off both architecture and culture in one blow… well, Beaubourg would then be the most audacious object and successful happening of the century.
>
> MAKE BEAUBOURG BUCKLE! A new revolutionary slogan. No need to torch it or to fight it; just go there! That's the best way to destroy it. Beaubourg's success is no mystery; people go there just for that. The fragility of the edifice exudes catastrophe, and they stampede it just to make it buckle.[47]

Thus it may be possible to regard Beaubourg as a large Lorenz attractor around which a complex network of dynamic urban, cultural,

46 Jean Baudrillard, 'The Beaubourg-Effect: Implosion and Deterrence', trans. Rosalind Krauss and Annette Michelson, *October* 18 (1981).

47 Ibid.

political, social, financial, theoretical matrices are drawn. Here one can see the resonance between W. Brian Arthur's observations that positive feedback plays a role in the dynamics of such complex, open systems and Baudrillard's commentary on Beaubourg;

> Thus this concave mirror: it's because they see the mass(es) inside it that the masses will be tempted to crowd in. It's a typical marketing device from which the whole ideology of transparency draws meaning. Or put another way, in presenting an idealised miniature model they hope to produce an accelerated gravitational pull, an automatic agglutination of culture as an automatic agglomeration of the masses. The process is the same: the nuclear chain reaction, or, the spectacular operation of white magic.[48]

Read in such a way the cultural centre, like the museum, appears as a strange social attractor which 'locks-in' culture around itself simulating that which looks and acts like culture but which for Baudrillard negates art and leads to the death of actual or real cultural activity:

> Thus everything is neutralised at the same time: Tinguely is embalmed in the museological institution and Beaubourg is trapped within its so called artistic contents.[49]

Central to Baudrillard's argument in the Beaubourg paper is the paradox that the cultural centre of Centre Pompidou negates culture by both absorbing and neutralising the creative and psychic potential of cultural activity. And this occurs due to the functioning of the system as a means by which particular social, cultural and political relationships are represented. Beaubourg is therefore, for Baudrillard, symptomatic of the hyper-reality of late 20th century life in which reality is masked and perverted by the semiological logic of what he refers to (throughout his work) as 'the third order of simulation', which is characteristic of late capitalism. In this third order, Debord's 'Society of the Spectacle'[50] has mutated into the spectacle which simulates and replaces society – we no longer have access to a basic reality but only consume a simulation of it.

48 Ibid.

49 Ibid.

50 Guy Debord, *The Society of the Spectacle*, trans. Ken Knabb (London: Rebel Press, 2004).

This argument was first developed by Baudrillard in *The System of Objects*, in which he argued that the function and utility values of objects such as furniture are rhetorical values generated by this system of objects and the dynamics of its internal system of signification. As such the system generates and conveniently pre-packages for mass consumption the iconography of use value which is attached to objects. Baudrillard claimed:

> Every object claims to be functional, just as every regime claims to be democratic. The term evokes all the virtues of modernity, yet it is perfectly ambiguous. With its reference to 'function' it suggests that the object fulfils itself in the precision of its relationship to the real world and to human needs. But as our analysis has shown, 'functional' in no way qualifies what is adapted to a goal, merely what is adapted to an order or system; functionality is the ability to become integrated into an overall scheme. An object's functionality is the very thing that enables it to transcend its main 'function' in the direction of a secondary one, to play a part, to become a combining element, an adjustable item, within a universal system of signs.
>
> The functional system is thus characterised, in a thoroughly ambiguous way, on the one hand by a transcendence of the traditional system under its three aspects – as the primary function of the object, as drives and primary needs, and as a set of symbolic relations between the two – and on the other hand by a simultaneous disavowal of these three mutually reinforcing aspects of the traditional system.[51]

As a development of this argument Baudrillard argued that at Beaubourg what is consumed is not culture, but rather that which is represented – through the functioning on the micro and local level of the Beaubourg system and at the macro level of social relations in the hyper-real 'Third Order of Simulation' as a *simulation* of culture. Thus, like the iconography of utility value, at Beaubourg all that can be accessed is the iconography of aesthetics which has been hijacked – and with this hijack comes cultural implosion which removes the potential for genuine cultural activity which might serve as an antidote or site of resistance to the simulacral world we circulate within and around.

51 Jean Baudrillard, *The System of Objects* (orig. French 1968), trans. James Benedict (New York: Verso, 1996).

As the Drewe and Myatt fraud illustrates the art gallery is not only defined by the system of its architectural fabric but is also constructed around a system of classification and cataloguing. Hence it is that the system of the archive might behave like other complex systems and demonstrate complex behaviour such as – locking-in, memory (an archive is, after all, a system of memory), irreversibility and be autocatalytic (that is self-perpetuating).

The case is used here to demonstrate how the art gallery is a system of distributed representation in so far as they observe, order, classify, differentiate and represent. This is demonstrated by the following example. At the centre of a rather obscure report on the 'International Symposium on the Conservation of Contemporary Art' the following words on the classification system of the museum are given with reference to a proposed Tate Gallery Installation File System (for 'Installation Art') In their absurdity they read like the scheme of a curator dreamt up by Borges:

> The file card design – The final layout and headings, have not, at the time of writing, been settled. The set of headings below summarises the kind of information that will be included. At some future date, it is hoped that the data will be computer-stored and storable in order to produce useful lists such as 'All works needing electric power'; 'All works with purpose-built storage facilities'; 'All paintings not requiring barriers.'[52]

Thus it is that if this system of archiving, differentiation and memory within the social system can be seen to be a system which is irreversible and which has become 'locked-in' then we could argue that the space of the gallery is the site of cultural inertia. It is the site, or attractor, around which the social system gravitates; and in doing so a particular configuration of social and cultural relationships also become 'locked-in'. Like Baudrillard, Robert Smithson too acknow-

52 Peter Wilson and Sandy Nairne, 'The Installation File – Conserving by Documenting', *International Symposium on the Conservation of Contemporary Art, 7–12 July 1980, Abstracts* (National Museums of Canada, 1980), pp. 16–19.

ledged the cultural inertia of the gallery identified as a process of locking-in:

> Museums are tombs, and it looks like everything is turning into a museum. Painting, sculpture and architecture are finished, but the art habit continues. Art settles into a stupendous inertia. Silence supplies the dominant chord. Bright colours conceal the abyss that holds the museum together. Every solid is a bit of clogged air or space. Things flatten or fade. The museum spreads its surfaces everywhere, and becomes an untitled collection of generalisations that immobilise the eye.[53]

4.3 Lock-in and Class

So far I have suggested that the concept of the complex system can be applied to the systems of the artworld with specific reference made to the art gallery/museum. This is in so far as there are observable patterns of 'lock-in' in both the archive system of the museum and the architectural system of the museum.

In conclusion a particular aspect of 'lock-in' within systems will be further extrapolated upon with a view to carrying it forward into the next, closing chapter. This is a theme which has already been mentioned in relation to lock-in and the effects of positive feedback. This theme suggests that cultural systems (like other complex systems) whilst being complex and unpredictable are also predisposed to inertia in so far as they lead to a certain stability being achieved within those systems. In Arthur's VHS/Betamax system a stability could be observed to emerge within the market. And in cultural systems this could be likened to the emergence of dominant cultural forms. An example of this inertia within a cultural system is illustrated by the suggestion that the YBAs were VHS art which cornered a market share over what I called 'Betamax art'. It is thus suggested that this systemic stability is manifested as a form of cultural inertia which, in turn, has implications within the larger social systems with which it interacts.

53 Robert Smithson, 'Some Void Thoughts on Museums', 1967, in Jack D. Flam (ed.), *Robert Smithson: The Collected Writings* (Berkeley: University of California Press, 1996), p. 42.

Bourdieu has written extensively about the role the gallery system plays in the re-enforcement of the social hierarchy of class structure. And while his approach can be critiqued as relying too heavily upon a particular conception of class distinction the central theme – that museum systems are a mechanism of cultural inertia or 'lock-in' – is one which I suggest can be recast in the vocabulary of a systems-approach.

In *Love of Art* Bourdieu attempted a thorough analysis of the public which visit European museums in order to:

> Devise a systematic survey of the European museum-going public, its social and educational characteristics, its attitudes to museums and its artistic preferences, as a *process of verification* aimed at confronting a coherent system of theoretical propositions with a coherent system of facts produced by – and not for – the hypothesis which it was necessary to validate.[54]

Bourdieu uses the questionable criteria of 'quality' of a work of art on display in museums which he grades on a scale from one to five and which has been assessed and 'evaluated by the panel of experts'[55] which he has used to establish a hierarchy of art works as displayed in European galleries. Despite the problems inherent in such a method which must rely upon the establishment of the quality of the artwork the hierarchy which is established can be seen to reflect a 'lock-in' in so far as that work which is perceived to be of a higher quality will attract more attention and visitors to the museum. As Bourdieu noted:

> Analysis showed that the majority of the museums' characteristics were strongly interrelated: disregarding the museums that receive less than 2,000 visitors a year (six museums, which are hard to reach and exhibit a small number of poorly presented works), it is apparent that the number of works displayed is strongly related to the number of visits (except in the case of certain large museums showing relatively few works but which are very famous or of great quality). The same applies to the fame and quality of the works (evaluated by

54 Pierre Bourdieu, Alain Darbel, Dominique Schnapper, *The Love of Art: European Art Museums and Their Public*, trans. Caroline Beattie and N. Merriman (Cambridge: Polity Press, 1991), p. 5.

55 These experts were constituted by 'a panel composed of five curators and art specialists selected from amongst all French museums, 123 art museums (containing paintings and sculptures)', ibid., p. 6.

> the panel of specialists), which tends to demonstrate that, as far as visiting rates are concerned, the 'official' hierarchy defined by the 'cultural authorities'.[56]

This provides another example of positive feedback loops within social systems with the art system gravitating around particular cultural forms. When the artwork is deemed to be of high quality, by the 'official hierarchy' it attracts visitors which, in turn as popular work attracts more visitors, attract more visitors; and so on. This leads to the lock-in of cultural inertia in so far as it leads to a dominance of the cultural forms which are identified and thus sanctioned by the 'official hierarchy' of the 'cultural authorities' and which subsequently cause people to be attracted to them.

When combined with a sociological critique of the function of museums based upon his own agenda of class distinction and analysis it is concluded that just as there is a 'lock-in' around certain museums so too there is a lock-in around the type of visitor to these museums:

> Museum visiting increases very strongly with increasing level of education and is almost exclusively the domain of the cultivated classes. The proportion of the different socio-economic categories in the French museum public is almost exactly the inverse of the proportion in the total population.[57]

Given that for Bourdieu increasing level of education is an index of class (in terms of 'socio-economic category') the art museum is 'locked-in' around the structural social order of bourgeois power relations and is both a function and instrument of such hierarchy. This is reflected in the education and the employment of visitors as Bourdieu noted:

> As the modal visitor to French museums holds a *baccalauréate* (55 per cent having at least this qualification), it is not surprising that the profile of the public, according to socio-economic category, should be very close to the profile of the population of French university students according to their social origin; the proportion of farmers and farm labourers in the public of French art museums is 1 per cent; that of industrial manual workers 4 per cent; craftworkers and trades

56 Ibid., p. 8.
57 Ibid., p. 14.

people 5 per cent; clerical staff and junior executives 23 per cent (of whom 5 per cent are primary school teachers), and upper classes 45 per cent.[58]

It is also noted that the time spent in the museum is a further index of both class and education: Thus the average time actually spent on a visit which can be taken to be an objective value given to the works on display, whatever the corresponding subjective experience might be (aesthetic pleasure, cultural goodwill, sense of duty or a mixture of all of these) increases in proportion to the amount of education received, from 22 minutes for working class visitors, to 35 minutes for middle class visitors and 47 minutes for upper class visitors.[59]

Education is an important indicator within Bourdieu's method of analysis because, he argues, it provides the means for one to decode the art on display and to therefore understand the systems of representation on display. It is thus the way in which certain social and cultural values and cultural forms perpetuate – in the preservation and transfer of cultural capital. From the systems theoretical perspective this transfer of 'cultural capital' is observed as a form of memory (irreversibility) within a social system; and that this irreversibility provides the means by which the social system might become 'locked-in' around a certain social order. As Bourdieu (et al) note both the quality and the legibility of a work of art is mediated by the values of the social system, and is hence both relative to that system and maintained by the memory of that system through, amongst other apparatuses, the gallery system and the system of education:

> Considered as symbolic goods, works of art only exist for those who have the means of appropriating them, that is, of deciphering them. An agents degree of artistic competence is measured by the degree to which he or she can master all the means of appropriation of works of art available at a given time: in other words, the interpretative schemata which are the condition of appropriation of artistic capital, that is, the condition of deciphering the works of art supplied to a given society at a given time.[60]

58 Ibid., p. 15.
59 Ibid., p. 38.
60 Ibid., p. 39.

And:

> The legibility of a work of art for a particular individual is a function of the distance between the more or less complex and sophisticated code demanded by the work, and the individual's competence, defined by the degree to which the social code, itself more or less complex and sophisticated.[61]

5. Conclusion

This conclusion comes as a rejoinder to the predominant theme of the chapter. It has, thus far, been suggested that whilst the museum system is complex an implication of this complexity might be that this system acts as an apparatus of cultural inertia within social systems. That is that the complexity of the museum system allows for, by the processes of irreversibility and the memory of the system which is manifested in the effects of positive feedback, for the preservation of particular cultural forms. And with the preservation of certain cultural forms at the expense of the exclusion of others certain social hierarchies, structures and power relations are also maintained. It would seem then that there might be the tendency to lay the blame for a particular configuration of social power relations upon 'the system'.

In response to this blaming of the system it is suggested that 'lock-in' within cultural systems rather than being seen as a purely negative phenomena which stifles all activity which it fails to validate by inclusion can also provide the conditions and potentialities for its own critique.

Robert Smithson recognised the potential for transgression and critique which the museum system provided:

> Well it seems to me that there is an attitude that tends toward McLuhanism, and this attitude would tend to see the museum as a null structure. But I think the nullity implied in the museum is actually one of its major assets, and this should be realised and accentuated.[62]

61 Ibid., p. 42–43.

62 'What is a Museum? A Dialogue between Allan Kaprow and Robert Smithson', (1967), in Flam, *Robert Smithson*, pp. 43–44.

In this respect Smithson's identification of the potential site of cultural transgression as the site of potential cultural subjugation can be likened to the Foucauldian invocation of 'power' as transformative as opposed to purely repressive. As Foucault argued:

> Where there is power, there is resistance, and yet, or rather consequently, this resistance is never in a position of exteriority in relation to power… [The existence of power relations] depends upon a multiplicity of points of resistance: these play the role of adversary, target, support, or handle in power relations. These points of resistance are present everywhere in the power network. Hence there is no single locus of great Refusal, no soul of revolt, source of all rebellions, or pure law of the revolutionary. Instead there is a plurality of resistances, each of them a special case: resistances that are possible, necessary, improbable; others that are spontaneous, savage, solitary, concerted, rampant, violent; still others that are quick to compromise, interested or sacrificial; by definition they can only exist in the strategic field of power relations.[63]

And hence it could be argued that rather than acting as a site in which the propagation of cultural activity is negated, as Baudrillard pessimistically feared, the cultural inertia of the museum system actually provides for the possibility of resistance, action and transgression. It is this potential for resistance to the power relations implicit in the systems of the gallery, art history and other cultural systems which is where the artistic impulse which has been called 'institutional critique' emerges. This is the artistic impulse which Robert Smithson identified, with foresight which was proved to be well-placed, as: 'The great issue, I think it will be the growing issue, of the seventies; the investigation of the apparatus the artist is threaded through.'[64]

Craig Owens discussed Smithson's comments further in his discussion of such an impulse in 'From Work to Frame, or Is There Life After "The Death of the Author"?' (based around the activities of such postmodern artists as Marcel Broodthaers, Daniel Buren, Michael Asher, Hans Haacke, Louise Lawler, Martha Rosler, Mary Kelly,

63 Michel Foucault, *The History of Sexuality; Vol. 1, an Introduction*, trans. Robert Hurley (Hammondsworth: Penguin Books, 1998), pp. 95–96.

64 'Conversation with Robert Smithson; edited by Bruce Kurtz', in Flam, *Robert Smithson*, pp. 262–263.

Allan Sekula, Robert Smithson et al.). Owens, like Smithson argued that it was imperative we engage with the:

> De facto social nature of artistic activity [in order that]... we ourselves are to employ rather than simply being employed by, the apparatus we all – 'lookers, buyers, dealers, makers' – are threaded through.[65]

This means looking to the systems around which art circulates.

Finally; this again raises the possibility of hanging a potentially huge ontological question mark over that which is not validated by being brought within the field of cultural production and which doesn't make it into the cannon. We thus return to the question that if practice and objects don't become institutionalised not only in galleries but also in art history then can they be called art – and if so on what terms?

I want to suggest therefore that discursive systems of art history, like galleries systems, are also locations around which 'lock-in' occurs. It is for this reason that we, who work in the artworld, should be aware that the processes and systems we are engaged in – in terms of everything we write and everything we say – are couched in terms of specific ideological positionings; which, whilst it might seem to be a small matter, can have large effects. The complex systems we operate in have an irreversible memory for which, despite our apparent lack of individual agency, we must take collective responsibility.

65 Craig Owens, *Beyond Recognition* (Berkeley: University of California Press, 1992), p. 136.

Chapter 6: Art Discourse as System

Introduction

The argument in this concluding chapter relies upon one assumption and advances two claims. The initial assumption, which will be subsequently demonstrated through three examples, is that art history is, like the art gallery, a complex system of observation, differentiation and representation. In the preceding chapter particular claims were made about the systems of preservation and display of art in so far as they displayed certain systemic principles. In this chapter I argue that the same patterns of systemic behaviour can be seen at work within the systems of art history.

The arguments concerning the assertion that art history behaves like a complex system operate by identifying and analysing key features of systemic behaviour in the system of art history, as they can be observed. These are that the system of art history is a distributed system of representation and that, as a result, the discursive systems of the art world (including art history)[1] are 'locked-in' around certain concepts (i.e. they display memory and inertia.)

In Chapter 4 the idea of the artwork as being distributed over a complex system was discussed via *Spiral Jetty*. When framed by a systems theoretical perspective, I argued, the dematerialisation of the singular artwork is best understood in terms of the ontological subordination of the singular artwork to the distributed systems of both the art gallery and the complex artwork itself.[2]

1 In other words, art history and art criticism are sub-systems of the system of art-discourse, which has particular behavioural patterns that will be discussed below.

2 A systems theoretical understanding of art will take the entirety of the relevant system as the fundamental object of analysis, rather than the singular artwork. Such an approach resonates with Luhmann's claims for a systems theoretical

Through the discussions of the institutional theory of art in Chapter 2 it was demonstrated that the artwork needs to be conceived as a node in a network of preservation and display. Also, through the example of *Spiral Jetty* it was argued that the system of distributed representation by which the artwork was observed was robust enough to stand the dissolution of certain aspects of the artwork, such as its submerging of the sculpture by The Great Salt Lake.

Such a dematerialisation raises a similar question to be asked of the discursive systems of art, such as art history. This question concerns the dematerialisation of the identity of the singular author in relation to their work; an author who is also implicated in the processes of distribution and dissolution.

In what follows I will initially approach this question by invoking Foucault's question: 'What is an Author?' I will then provide an answer to this question that is framed from within an art historical as well as from a systems theoretical perspective. Let me begin with one of Foucault's observations:

> Yet, what of a context that questions the concept of a work? What, in short, is the strange unit designated by the term, work? What is necessary to its composition, if a work is not something written by a person called an 'author'.[3]

Foucault treated the author not in terms of the individual creative subject but instead as 'a function of discourse'. I will argue below that such an argument can be re-stated from a systems theoretical perspective. I propose that the author (or artist) can also, along with the

sociology, which analyses not individual human subjects but social systems as fundamental objects of analysis. In Luhmann's words: 'If one views human beings as part of the environment of society (instead of as part of society itself), this changes the premises of all the traditional questions, including those of classical humanism. It does not mean that the human being is estimated as less important than traditionally. Anyone who thinks so (and such an understanding either explicitly or implicitly underlies all polemics against this proposal) has not understood the paradigm change in systems theory.' Niklas Luhmann, *Social Systems*, trans. John Bednarz and Dirk Baecker (Stanford: Stanford University Press, 1995), p. 212.

3 Michel Foucault, 'What is an Author' (1969), in Donald Preziosi (ed.), *The Art of Art History* (Oxford: Oxford University Press, 1998), p. 301.

gallery and the artwork, be conceived as a function of a system rather than as a unique originating source. From this position I argued that, like the dematerialised *Spiral Jetty*, the author/artist need not be present in order to continue to be represented by the distributed systems of the artworld.

In response to Foucault's question the role of author is occupied by the artist. In the following, I will discuss three examples of artists who are, in different senses of the word, absent. They, like the postmodern art object, dematerialised. Nevertheless these artists have been re-presented by the distributed systems of art discourse (including art history) as the originating 'authors' of their work. And this tells us something of the ways in which these discursive systems function. These three 'authors' are:

1. Rembrandt: an artist who lived in the past and still exists in the artworld system;

2. Nat Tate: an artist who never lived and did not exist in the artworld system;

3. Hank Herron: an artist who never lived yet who did have an existence in the artworld system.

Contrary to the standard view that these three artists constitute exceptions, or even distortions, of the normal discourse on art I propose that they are instead paradigmatic for the general behaviour of the systems of art discourse.

1. Art History and the Author Function

Vasari famously put the individual artist at the centre of his history of the 'Lives of the Artists'. In doing so he firmly linked artworks with the actual manifestation (and material mediation) of the artists' individual subjectivity. Such prioritisation of the artistic subject was subsequently dissolved in the various attempts since to provide a different set of criteria for historical reflection. These include analyses such as Wölfflin's 'Art History Without Names', which he explicitly conceived as an attempt to distance himself from overly biographical considerations. The desire to differentiate a critical art history from questions of biography in favour of addressing more general cultural questions can also be seen in Riegl's claim that the artistic 'genius' does not, as is often thought, 'stand outside' the cultural systems within which art operates.[4] It is also revisited in Panofsky's iconological archaeology of social systems through the systems of artistic representation.

Similarly, in her defence of a critical discipline of art history that is grounded in formalism, Rosalind Krauss launched a critique of 'art history as a history of the proper name', a biographical art history that singles out particular artists as the central focus for critical attention:

> It is as though the shifting, changing sands of visual polysemy, of multiple meanings and regroupings, have made us intolerably nervous, so that we may wish to find a bedrock of sense.[5]

Many writers find this 'bedrock of sense' in the identity of the author/artist who provides the key, coherence or origin to a body of work.

The identification of the author as the single and unique origin for the work of art faced two major problems in its confrontation with the modernist work of art. The first challenge arises in the light of the pronounced autonomy of the modernist art object. The proclamation that the modernist art object can speak on its own terms, as defended by Greenberg,[6] means that qualities extraneous to the 'purely plastic or abstract qualities of the work of art' are negated; such terms might

4 'Geniuses do not stand outside their national tradition, they are an integral part of it… The great artist, even the genius, is nothing but the executor, though the most perfect executor, the supreme fulfilment, of the *Kunstwollen* of his nation and age.' Alois Riegl, cited by Michael Ann Holly in *Panofsky and the Foundations of Art History* (Ithaca: Cornell University Press, 1984), p. 75.

5 Rosalind Krauss, 'In the Name of Picasso', *October* 16 (Spring, 1981), pp. 5–22.

6 'The purely plastic or abstract qualities of the work of art are the only ones that count. Emphasise the medium and its difficulties, and at once the purely plastic, the proper, values of visual art come to the fore.' Clement Greenberg, 'Towards a Newer Laocoon', (1940), in Charles Harrison and Paul Wood (eds.), *Art in Theory 1900–1990* (Oxford: Blackwell, 1992), p. 555.

include not only the language of the art's iconography but also the biographical details connected to the author.

The second challenge to the 'Art History of the Proper Name' comes in the aftermath of the demise in faith in the modernist art object. The subsequent disappearance of the autonomous art object did not see the re-instatement of the author/artist as the prime origin of the work as is demonstrated by the post-structural claims of the 'Death of the Author'. Within the field of art this disappearance occurred when the modernist work of art was questioned from within both artistic and art historical practice in favour of diverting attention to the various systems of support. Such a critique is exemplified by the claim of artistic relativism Rauschenberg made by sending the telegram 'This is a portrait of Iris Clert if I say so' (1962) as a proxy for an absent object (and by which he, the artist himself, also distanced himself from the work).

Despite the artistic statement of post-modernism, such as Rauschenberg's, it is in the face of the loss of the modernist work of art that the attractiveness of the 'Art History of the Proper Name' might potentially be seen to increase. This is by using the concept of the author/artist as an anchor of attention and as a site around which a discursive system can become locked-in. This is to say that, given the loss of a discrete object, the proper name of the artist can function in a system of representation as a stand-in for the dematerialised art object; as it does in Rauschenberg's telegraph. However what also needs to be acknowledged is that this statement of Rauschenberg's does not stand as a reclamation of mastery of the artist over the act of representation. Whilst it makes any easy differentiation between art object, artist and environment impossible it is also true that the work can only function because the figure of the artist depends upon the same continuum as the artwork itself. This continuum is the system of distributed representation within which object, artist and artworld environment are all incidents of representation.

1.1 In the Name of Rembrandt

The problem which the 'author function' presents to the discursive systems of art history is exemplified in the first of the three artists I will look at in this chapter, namely the work of Rembrandt.

The large body of Rembrandt's work presents several problems for the systems of art history. Central to these problems is the very persona or 'name' of Rembrandt itself. Clearly the personality of Rembrandt (like that of Michelangelo or Caravaggio for example) is inextricably bound up with the body of work referred to as 'Rembrandt'. This holds to such an extent that this art historical personality (as opposed to the biographical or biological one) dominates much of the discourse. The most acute illustration of this is provided by the series of paintings for which he is arguably most famous. These are the self-portraits which he made throughout his life. With these portraits the person and artist 'Rembrandt' has himself become something of the work of art; his personality has thus explicitly become the object of art historical discourse. As Svetlana Alpers observed, this raises certain questions:

> Individual uniqueness, monetary worth, aesthetic aura – the name of Rembrandt summons up different but nevertheless convergent notions of value. And to what precisely does the name itself refer? *Rembrandt* as it is commonly deployed tends to elide the works with the man. Is all of this a recent invention, a prejudice of the nineteenth century to which we are heir? Or is it appropriate to Rembrandt's own making a marketing of his own works that they should offer an example of that relationship between the production and valuing of a person, property, and aesthetic worth that lies so deep in our culture.[7]

The corpus collected under the name 'Rembrandt' thus provides us with an excellent example of the 'Art History of the Proper Name', which Krauss had critiqued. Yet, as I will further demonstrate, this is an example that at the same time also subverts the operations of such art historical strategies.

7 Svetlana Alpers, *Rembrandt's Enterprise* (London: Thames and Hudson, 1988), p. 3.

The way in which the system of art history has dealt with such problems is of particular interest to my project as it concerns the identification of systems within art historical practice. For it is in these negotiations that the very operations of those systems of art history become obvious and are thus made available for scrutiny. In the following section I will use the work of the Rembrandt Research Project to show that within the systems of art history the way the proper name 'Rembrandt' functions as a descriptor and attractor within a complex representative system. In short, for the whole body of work which makes up the corpus of 'Rembrandt' and is archived under that name, 'Rembrandt' serves as an organising principle rather than as a signifier for a single originating source.

The Rembrandt Research Project (RRP) was established in 1968 with a grant from the Netherlands Organization for the Advancement of Pure Research. It was born from a perceived need amongst scholars to scrutinise and possibly reduce the large number of works (which at one stage had reached over a thousand) attributed to the single source of Rembrandt. For there were clearly numerous works that did not originate from (the historical person) Rembrandt's own hand. As the RRP identified, 'by the 1960s it was difficult for an impartial eye to accept all the works currently attributed to Rembrandt as being by a single artist.'[8]

8 Rembrandt Research Project, *A Corpus of Rembrandt Paintings I* (Stichting Foundation/Rembrandt Research Project, 1982), p. I. The activities of the RRP engage in issues similar to those faced by the Warhol Art Authentication board, a body which exists to secure the authenticity of Andy Warhol's work. A controversy concerning the board's activities was initiated when film producer Joe Simon tried to sell a Warhol self-portrait for 2 million dollars. Apparently, although the image was not signed, it did bear a stamp of authenticity from Fred Hughes, Warhol's business manager and also had supporting paperwork. However, when the picture was submitted to the board that was to guarantee its authenticity it was rejected. Given the nature of Warhol's practice this issue is both sensitive and complex. So far about 15% of the artworks re-appraised by the board have had their authenticity removed. It was reported in *The Telegraph* that: 'Ron Spencer, the lawyer for the authentication board, said that its role was to ascertain the "intent" of the artist [... and] that even if a work were printed by a number of other people the board would still classify it as genuine. If Warhol conceived the idea and he then directed someone else to prepare a silk-

One result of this process of differentiation that arose from the RRP's investigations was the establishment of three categories for further classification of 'Rembrandts':

First, those works identified beyond all reasonable doubt as being by Rembrandt himself.

Second, those of uncertain authorship. (The RRP acknowledged that this second category was necessary but also pointed out that given its ambiguity this category should be kept as small as possible).

Third, those works identified as *not* produced by Rembrandt. These, previously accepted as authentic, works included paintings by his studio or 'after Rembrandts' as well as fakes and pastiches. Examples of these types of work include two works in The Hague now officially accepted as non-Rembrandts, *The Polish Rider* and *David and Saul*, as well as *Man With the Golden Helmet* (Berlin).

The first stage of the RRP (between 1968–1972) involved collating all the information that existed for all works that were attributed to Rembrandt. This was done in the form of a literal description of each work and an accompanying photographic record. This meant that the RRP had a established and definitive body of works which were, at the time, recognised as 'Rembrandts'. In the second stage of the RRP's modus operandi this larger set of 'Rembrandts' was reduced to a much smaller number of works comprised of all those pieces that be said to consistently display having clear similarities to one another.

The initial information used by the RRP was taken from an established and recognised art historical archive. This was the catalogue of Rembrandt's work compiled by Bredius in 1935. The assumption was that this would, at the very least, contain all the works by Rembrandt in existence along with other work. In total this body of work of unproven provenance was estimated as comprising about a thousand paintings. It should be noted that the choice of Bredius' catalogue already meant the exclusion of an even larger body of work

screen, supervised the process of production and, in effect, signed off on it, as long as he said, *That's good, that's what I wanted*, Warhol created that work.' Quoted in Catherine Milner, 'Collectors insist "our Warhols are genuine"', *The Telegraph* (26/10/2003). See also the originating article by Michael Shnayerson in *Vanity Fair* (November, 2003).

based on 19th-century compilations that included works of dubious provenance which had nonetheless been identified, at some point, as 'Rembrandt'. By taking the task of defining and purifying Rembrandt's oeuvre as the guiding principle and ethos of their method the RRP have so far been able to reduce the number of what is generally accepted as Rembrandt's work to below four hundred pieces.[9]

Starting from this central question of attribution the members of the RRP then sought to establish working strategies that could accommodate and address the particular problems that the Rembrandt corpus presented.

The first major problem with the corpus was that what was generally understood to look like Rembrandt's work was (and still is) both highly distinctive and extremely well-known. As a result of that fame and recognition, Rembrandt's originals have been much copied and emulated in acts of homage, pastiche and fraud. The fact that there are so many works referred to as 'Rembrandt' which, however, cannot beyond reasonable doubt be attributed to the hand of Rembrandt Harmensz van Rijn of Leiden who died in 1669 (for example the *Flayed Ox* at Kelvingrove Museum in Glasgow which until relatively recently was wrongly attributed to Rembrandt) makes it very clear how problematic the attribution of works of art to a single individual artist/author is.

Secondly, the case study of Rembrandt and the RRP further highlights the problem of attributing works of art to specific historical individuals with regards to the issue of the unique autographed work. Common to working practice in 17th-century Holland, Rembrandt worked with a studio of apprentices and assistants; in fact he had over fifty students and assistants working for him during his working life.[10] This means that many hands may have worked on a work which was subsequently identified as an 'original' Rembrandt.

However attribution is a complex issue. On the one hand, Rosenberg et al. observed that:

9 In collating their results the Rembrandt Research Project produced *A Corpus of Rembrandt Paintings Vol. I-V* (Stichting Foundation/RRP, 1982), which they hoped would become the authoritative work of attribution.

10 Alpers, *Rembrandt's Enterprise*, p. 59.

> In the Night Watch and the Conspiracy of Julius Civilis – large pictures for which he could have used help – there is no evidence that a hand other than the original touched the original work.[11]

On the other hand they observed that Rembrandt's studio system was markedly different from that of, for example, Rubens. Rubens undertook large scale decorative projects (such as the decorations for the Luxembourg Palace, 1622) and as a result would have used a whole team of assistants as a normal working practice. The very fact that Rembrandt does not seem to have worked on such collaborative projects might be taken to suggest that in his case single works might indeed be attributed to a single, originating source.

However it is also known that Rembrandt worked closely with his pupils and that he often altered drawings they had made. Thus, consequently his hand may have been at work in a number of pieces which he neither started nor finished but which were instead executed in the main by students and colleagues. Further, as Rosenberg et al. claimed:

> It is difficult to determine how many Rembrandt school pictures are based upon the master's own inventions, but the number is probably high, and in a few cases a definite connection can be established.[12]

The examples they cite in order to give an impression of the way in which the studio operated include an engraving after the painting *Anna and the Blind Tobit* (1630, National Gallery, London). It is signed 'Rembr. van Rijn fecit' despite the fact that the original was identified as having been painted by Gerrit Dou. Dou was a member of Rembrandt's studio and there is evidence that Rembrandt did add some cursory last minute touches to the painting. Those works which had been updated by the master were identified at the time as being 're-touched by Rembrandt'.[13] These include the *Sacrifice of Isaac*, attributed to Govert Flinck, (the Pinakothek in Munich) which is based

11 Jakob Rosenberg, Seymour Slive, Engelbert Kuile, *Dutch Art and Architecture 1600–1800* (New Haven: Yale University Press, 1966), p. 142.

12 Ibid.

13 'Van Rembrandt geretukeert', from the 1656 inventory of Rembrandt's effects, cited in Rosenberg et al. (1966).

upon the master's treatment of the same subject matter (1635, Leningrad and a drawing in the British Museum). According to Rosenberg et al.:

> The inscription in Rembrandt's hand at the bottom of the Munich (*Rembrandt verandert En overgeschildert. 1636*) announces that the master was not satisfied with his pupil's results, and that he changed and retouched the work.[14]

The issue of Rembrandt's signatures, especially with regards to its appearance on work that didn't otherwise conform to the criteria of being 'Rembrandt' is one which the RRP diverted particular attention to as part of their attempts to provide empirical evidence for their attribution of works. They wrote:

> In some instances they do come so close in form to that of an autograph signature that one wonders whether the artist may not also have put his signature on paintings produced, in his workshop and under his supervision, by others.[15]

The above quotation is taken from an article by the RRP on the unfinished painting, *Self-Portrait* from 1642 in the Royal Collection. The self-portrait was originally derived from another painting which Rembrandt began and then abandoned but which was later updated and thus 'finished'. The RRP have argued that such alterations provide further evidence of the work's originality since 'traces of other paintings are found in approximately one third of Rembrandt's accepted self-portraits.'[16]

This brief sketch of some of the working methods of the RRP leads me onto a more detailed description of how they conduct the process of differentiation and attribution. I propose that this process demonstrates a systems principle – namely that of distributed representation. But before I present this account in systems theoretical

14 Rosenberg et al., *Dutch Art and Architecture*, p. 142.

15 Rembrandt Research Project, 'A Descriptive Survey of the Signatures', *A Corpus of Rembrandt Paintings I*, p. 59.

16 Ernst van de Wetering and Paul Broekhoff, 'New Directions in the Rembrandt Research Project, Part I: the 1642 Self-Portrait in the Royal Collection', *Burlington Magazine*, vol. 138, no.1 (1996), pp. 174–180.

terms I first will outline in further detail the most relevant features of the working methods of the RRP.

The working method of the Rembrandt Research Project is based upon thorough empirical research supported by technical and scientific methods of material analysis. In doing so the RRP thus return to the materiality of the art object as the central focus for their interpretive method. The technical means that the Rembrandt Research Project have at their disposal to facilitate the process of attribution include: dendrochonology (which is the dating of the wood of support on painted panels); x-ray imaging (to see earlier versions of the image including pencil sketching and so-forth); infra-red photography; microscopy (to identify micro-details of the support); and autoradiography (in which a beam of neutrons is fired at the painting to help identify aspects of the layers of paint.) The rationale behind this was to seek to establish grounds for authenticity on material terms.

However although the project went to great lengths to provide rigorous empirical evidence for their attribution ultimately, the process of differentiation was not a purely technical one. Whilst the description of the paintings was primarily grounded in purely materialist terms (in terms of paint, canvas, frame etc.) which was practiced as being the most basic and fundamental act of investigation and differentiation this was, in turn, mediated by the interpretative act of art historical representation. It is a notable feature of the RRP that it is composed not of scientists but of art historians. Therefore, as expressed by members of the RRP the interpretative act of historical evaluation does takes precedence over pure material empiricism:

> Apart from elementary information on the materials used, we have not attempted a systematic study of pigments, media, drying agents, dilatants etc; such studies may yield further specific technical information as analytical methods become more refined, though it remains to be seen whether the results will help solve problems of attribution. We have, rather, selected such information as can clarify the stratified structure of the painting as it results from the actual painting procedure.[17]

17 Rembrandt Research Project, *A Corpus of Rembrandt Paintings I*, p. xi.

This refusal to rely upon purely technical data in the identification of the works was to allow for the foregrounding of particular interpretative reasons of attribution based on stylistic and qualitative evaluations on the part of the art historians.

According to the RRP, the difference between a purely materialist and an art historical investigation and historical reconstruction is:

> Provided by the differences in the sort of questions asked and the working method adopted by a scientist and an art historian, even when they approach the work of art as a shared object of study. Each is conditioned by the traditions of his own discipline. Without being unfair to either, we might perhaps say that the scientist arrives at his interpretation from relatively fragmentary and, of itself, unstructured information relating to the physical make-up of the work of art, while the art historian is concerned mainly with the stylistic interpretation of the picture and its execution. Their common frame of reference ought to be an understanding based on source studies, of the craft that governed artistic practice.[18]

It was also noted by the authors of the project that materialist accounts of the work in terms of attribution had already been well developed in Rembrandt studies, but that this was limited in so far as: 'this is only seldom clearly related to what the art historian is seeking.'[19] What the art historian is seeking, in this case, is the process of historical reconstruction. Thus the attribution of works to the Rembrandt corpus requires the interpretative act of the art historian (or historians in this case) to establish a coherency between works beyond that of mere materiality. This coherency can be identified as an organising principle in the systems of both the Rembrandt Corpus and larger artworld systems (such as art historical discourse, or the gallery system, and so forth). These grounds of authenticity are based on consistency amongst the accepted pieces of the corpus (within the system of Rembrandt). And if anything doesn't correspond to the organising principles then it is be deemed as inauthentic and hence not a 'Rembrandt'. This coherency is, in turn, achieved through the act of historical interpretation which thus requires the imposition of a narrative between the work and the 'Rembrandt' corpus. I propose that this nar-

18 Ibid., p. xii.
19 Ibid.

rative can be seen, from a systems theoretical perspective, as a system which observes, represents and hence differentiates some works as belonging to the Rembrandt oeuvre and others as not; it does so around the organising principle of *paintings that look like Rembrandts.*

As it can safely be assumed that Rembrandt and his shop assistants used the same materials and procedures, it was inevitable that the members of the Rembrandt Research Project, like their predecessors, had to rely on connoisseurship to distinguish autograph from workshop productions.[20]

Thus, in the attribution of identity the project invoked the working method of Morelli[21] who claimed that a process of attribution should be informed by the connoisseur's eye which was tuned to the smallest details of the work. In doing so the connoisseur would be able to identify the involuntary stylistic ticks and unconscious habits of the painter's style in otherwise similar paintings. In the case of Rembrandt however the analysis of attribution of works via brushstrokes was considered insufficient by the members of the RRP. This was for two reasons.

First, unlike the art of the Renaissance, which Morelli concentrated on, Rembrandt did often not work on the incidental details of his paintings but instead left such matters to be completed by his studio.

Second, Rembrandt's brushwork was (and still is) highly distinctive. For this reason it is one of the characteristics of his art that was most appropriated by students and members of the studio, copied by contemporaries and those whom he influenced and parodied by forgers. This makes subsequent attributions based on attributes of his handling of pens and brushes highly problematic.

20 Wetering/Broekhoff, 'New Directions', pp. 174–180.

21 See Domenico Morelli, *Italian Painters* (1890), in which Morelli, in the narrative guise of an elderly Italian traveller whom the author meets relates his thesis on the nature of a correct art historical method founded upon close attention to detail and informed connoisseurship. 'A true knowledge of art is only to attained by a continuous and untiring study of form and technique, that no one should venture into the domain of the history of art without first being an art connoisseur.' Quoted in Eric Fernie (ed.), *Art History and its Methods* (London: Phaidon, 1995) p. 110.

It was because of the problems in using techniques of the artist's hand, such as brushstrokes, that the Rembrandt Research Project required a different conceptual organising principle by which to differentiate the *paintings that looked like Rembrandt* from other works of art. In doing so they turned, ultimately, to questions of 'style'. This was complicated procedure because the Rembrandt Corpus had become locked-in around the style of Rembrandt; hence the need for the project's work in first instance.[22]

In summary, the working method of the RRP was an intricate process of differentiating those paintings from the larger corpus that appeared to be sufficiently 'Rembrandt-esque'[23] and not an attempt to identify a *single* authorial voice manifest in all of them. To the contrary, RRP tried to determine a set of *multiple* criteria that all the works within the oeuvre of 'Rembrandt' would display.

Thus a 'Rembrandt' is not defined by virtue of an external reference which the work signifies; i.e. Rembrandt the artist. The process of attribution and differentiation instead functions according to the criterion of internal coherence within the system 'Rembrandt'.

22 'The process of illegitimate accretion to the oeuvre, which took place in the 18th and even as early as the 17th century, can be glimpsed from the prints put out in those years and purporting to reproduce paintings by Rembrandt. When John Smith published the first catalogue of the paintings, in 1836, his work inevitably reflected a corrupted tradition and consequently gave a distorted view. Eduard Kolloff (1854) and Carel Vosmaer (1868) deserve credit for bringing some kind of order into chaos, as Scheltema had done for biography; but it was particularly the young Wilhelm Bode who, in the 1880s, produced a corrected image of Rembrandt's work, especially that from the early years.' From: Rembrandt Research Project, *A Corpus of Rembrandt Paintings I*, p. lx. Other names which are cited as providing the foundations for the body of work accepted as 'Rembrandt' at the time of the beginning of the RRP include: Hofsted de Groot (1915), Hans Sedlmayer (1897–1905), Kurt Bauch (1966), Alfred von Wurzbach (1911), John van Dyke (1923).

23 Broos and Alpers identified three broad types of 'Rembrandtesque' paintings which show his influence both materially and conceptually. Alpers classifies them thus: '(1) Study heads taken home by pupils, (2) copies after Rembrandt and (3) paintings by the assistants sometimes perhaps even signed and sold as Rembrandts.' See Alpers, *Rembrandt's Enterprise*, p. 125. Ben Broos, 'Fame Shared is Fame Doubled', in *Rembrandt Impact of a Genius* (Amsterdam: K. and V. Waterman, 1983), p. 41.

And it is due to this internal coherence that a unity emerges among the works of the Rembrandt system regardless of whether they were painted by one artist or not.

Paradoxically the focus on style could also be interpreted as an (if only unintentional) return to the author in art history. As Bal concluded on the work of the Rembrandt Research Project:

> True, some Rembrandt critics (the members of the Rembrandt Research Project are among them...), voluntarily question the quality of some of the works, and even take pride in their demythologising attitude. Yet the very fact that these aesthetic criticisms lead to the rejection of the authenticity of the works as *Rembrandt* provides evidence of the basically unchallenged status of both author as genius and of aesthetic quality as determinable. Hence these acts of judgement rest on the possibility to delimit *high art* from other art. An ever *higher* purer Rembrandt is thus safeguarded.[24]

And:

> That many of these works are now in the process of being disavowed as part of the Rembrandt does not bother me; it rather supports my argument that *Rembrandt* is a cultural text rather than a historical reality. *Rembrandt* constitutes these works and the response to them, responses that range from their mistaken attribution to attempts to challenge their authenticity on the basis of a holistic and elitist concept of authorship.[25]

However although the RRP take the question of authorship as a starting point, their conclusions demonstrate that ultimately rather than being defined purely in terms of either materiality, or the individuality of the artist *Rembrandt* is instead a product of the discursive systems of the artworld. This highlights the role which these systems play in the processes of art historical observation, representation and differentiation.

24 Mieke Bal, *Reading Rembrandt* (Cambridge University Press, 1994), p. 7.
25 Ibid., p. 8.

1.2 In the Name of Nat Tate

On March 31st 1998 a party was held in Jeff Koons' SoHo studio in New York. Amongst the art dealers, writers, agents and other usual figures who attend such events were several high profile guests including the artists Julian Schnabel and Frank Stella, the author Paul Auster and David Bowie.

The party had been arranged by Bowie's publishing company (*21 Publishing*) to celebrate the publication of William Boyd's biography of Nat Tate, an 'American Artist' who had died in 1960. Bowie read several extracts from the book to the audience. Boyd's book provided a brief biographical sketch of this would-be mythical figure from the history of American Modern art. The story was illustrated by photographic images of the time, some documenting Tate and others key figures from the story, and three of Tate's own works. The works reproduced showed two drawings from Tate's series of over 200 images said to be inspired by Hart Crane's epic poem, *The Bridge.* The third image was an abstract oil painting called *Portrait of K*, reputed to have been painted in 1958 in a drunken frenzy; the style appeared to be reminiscent of Abstract Expressionism, although in the text the work was referred to as Abstract Impressionism.[26]

Nat Tate the man, however, never existed. The whole event had been part of an intended joke to play on the culterati of New York. The scheduling of the party on the day before April Fool's Day had been a deliberate attempt to place the story in the next mornings papers. The intention had been to coax those at the party into claiming to have heard of the (fictional) figure of Tate and for the papers to replicate this mistake the following morning. Karen Wright, a co-publisher of the book said later:

26 'Abstract Impressionism' was the description which Elaine de Kooning had used to describe the early colourful painterly abstraction of Phillip Guston. See Elaine de Kooning, 'Subject; What, How or Who?' *Artnews* (April 1955) cited in Irving Sandler, *The New York School* (New York: Harper and Row, 1978), p. 55.

> Part of it was we were very amused that people kept saying, 'yes I've heard of him'. There was a willingness not to appear foolish. No one wants to admit that they've never heard of him. But critics are too proud to admit that.[27]

However, as *The Times* reporter in New York related, the joke fell a little flat:

> But the hoax did not come off. The party went with a swing, but not one word appeared in any of the newspapers the following day. Or even the day after. Clearly, no-one had taken it seriously. Reports have now filtered through from London to the effect that *The Independent* carried a front-page story about how the novelist 'fooled the US art world'. The *Evening Standard* followed suit. But the question everyone here is asking is: What hoax?[28]

The Independent was indeed the first paper to break the story a week late on April 7th, although *The Telegraph* had serialised some of Boyd's book the previous week. This was followed up two days later with another story under the congratulatory headline: 'London Party-goers Revel in Hoax that took in New York's Smartest'[29] under which it was reported that:

> One branch of Waterstone's bookshop reported yesterday that it had had numerous requests for the 'book about the American artist written about in *The Independent*'.[30]

All things considered the joke of Nat Tate appears not to have fooled a gullible artworld into believing in the existence of an artist who never lived. But, equally, neither did it appear to be a totally incredulous story. There was something about Nat Tate that did seem at least in some sense plausible. One only needs to compare his story with that of the real-life 'lost' Abstract Expressionist John Schueler, a painter under the tutelage of Reinhardt, Rothko and Guston, who was

27 David Lister, 'How a British Novelist Fooled the US Art World', *The Independent* (April 7th, 1998), pp. 15 f.

28 Tunku Varadarajan, 'British Joke Falls Flat in New York', *The Times* (April 8th, 1998), p.11.

29 David Lister, 'London Partygoers Revel in Hoax that took in New York's Smartest', *The Independent* (April 9th, 1998).

30 Ibid.

also a jazz musician and who was described in the very recent catalogue of his first one man show in the UK (2003) as:

> A well established member of the younger generation of the New York School by the time Leo Castelli, an influential art dealer, opened his first gallery with a Schueler exhibition in 1957. The authors of the catalogue even go on to say that had Schueler stayed in New York, his career would have progressed in step with his more famous friends.[31]

Schueler instead chose to move to Mallaig and obscurity on the West coast of Scotland meaning that his work, like that of Nat Tate, drifted out of the artworld system.

The life story of Nat Tate used a series of motifs common to the typical biographies of art historical discourse. They occur with regularity within the system(s) of art historical biography and thus become representational tropes which constitute and demonstrate a systems principle. In other words such narrative tropes are organising principles around which the systems of biographical art history writing are 'locked-in'.

What follows are six such principles to be found in the narrative of Nat Tate's life. It is easy to illustrate how typical these principles are to art historical biographies in general. This is illustrated by the fact that some of these six principles are also applicable to the real-life figure of Schueler (with the exceptions that Schueler died early and was an orphan).

First, Nat Tate was a solitary figure who, like Van Gogh or Rembrandt did not receive the fame which he deserved in his own lifetime heightening the tragedy of his story.

Second, he was firmly embedded in the art world milieu of the Abstract Expressionists, supposedly having been tutored by Hoffman, written about by Greenberg (who called his work 'promising, oddly disturbing…'), slept with by Peggy Guggenheim and befriended by Jackson Pollock. He was thus a suitable candidate to bask, posthumously, in their reflected glory.

31 *To the North: Paintings by John Schueler*, brochure to the Exhibition at City Art Centre, Edinburgh (5th July –27th September, 2003).

Third, like Jackson Pollock, Tate was the existential hero of his own drama who was tortured by the twin torments of his genius and his alcoholism. One only needs to see Ed Harris' hagiographic film *Pollock* (2001) to see how pervasive this notion is.

Fourth, he witnessed a moment of artistic epiphany. In 1959 Tate was said to have visited Braque in Normandy. A photograph in the book corroborated this meeting at which Tate was reported to have seen Braque reworking a much older canvas. It was this, Boyd reports, which lead him to his final artistic statements by attempting to radically rework all the previous work which he could get his hands on:

> Throughout December 1959 it is clear that Nat Tate tried to buy back or asked to be allowed to 'rework' as many as possible of his paintings as were in public hands. There is no reason to doubt that he was sincere in this regard, that it was not, in Mountstuart's words, 'little short of theft'. Nat Tate had seen Braque at work, had witnessed his tireless and dogged perfection at first hand, and it is entirely conceivable he was inspired by Braque's example. In any event, he locked himself away in his Windrose studio and worked uninterrupted through the holiday season and into the early days of 1960.[32]

Tate's reworking of his art actually meant the destruction of his paintings; and this suggests a fifth narrative trope at work in the story of Nat Tate.

Fifth, conveniently practically none of Tate's work was left after his suicide. The story of Tate as an heroic, 'lost' artist is undoubtedly helped by this as there was very little material evidence that could have been used as evidence to disprove his 'genius'. This was a trick of artistic re-invention which can be seen in the oeuvres of Malevich and Hans Hoffman whose reputations benefited from being their early work having been fortuitously destroyed[33] and their subsequent reputations having been generated in the absence of paintings dating prior to the mature phases of their work.

32 William Boyd, *Nat Tate An American Artist: 1928–1960* (Sidney: 21 Publishing Ltd., 1998), p. 56.

33 This process of artistic re-invention has been revisited at least twice in the last forty years: by John Baldessari's Cremation Project (in which he burned his previous work) and Michael Landy's *Breakdown* (2001).

Sixth, Tate took his own life by leaping from the Staten Island Ferry in New York on January 12th, 1960. This motif of the suicidal artist can be seen in the lives of Gorky and Rothko and has a specific narrative similarity to the poet Hart Crane's drowning after jumping off a steamboat in the Caribbean in 1932. That Tate was an orphan who was only 32 years old when he died only serves to heighten this tragedy.

Yet despite displaying all of the above persuasive and convincing biographical characteristics, Nat Tate never gained credibility. However, that the fact that Tate did not exist as a 'real-life' person was not the reason that his story lacked plausibility. Rather it was because his story never gained the required credibility within the systems of art history. His story was never convincing enough for him to 'exist'.

The third and final example of authorship within the discursive systems of the art world will demonstrate what this credibility entails. As will be shown, the artist can indeed have a historical 'existence' in the systems of art discourse. The example of the artist Hank Herron demonstrates once more that the discursive systems of the art world and more specifically the system(s) of art history are best understood as distributed systems of representation. In fact it is their distributed nature that makes them robust enough to sustain themselves in the absence of the artist (as a person who lived) without losing the *figure* of the artist/author around which a body of work may be 'locked-in'.

1.3 In the Name of Hank Herron

Hank Herron was an artist who never lived. He was written about by a critic Cheryl Bernstein who also never lived. But both existed. As will become clear from what follows, however, in terms of the discursive systems of art both Herron and Bernstein had a particular type of existence in the discursive systems of art. For unlike Nat Tate both Bernstein and Herron gained enough credibility for the discursive systems of the art world to 'lock-in' around them. They became, like Rembrandt, representations within art discourse.

Both Herron and Bernstein were invented by the Feminist critic and author Carol Duncan (with her husband Andrew Duncan) in 1970.

In her brief written introduction to the anthologised versions of the two articles that appeared under Bernstein's name ('The Fake as More' and 'Performance as News'),[34] Duncan explains that Bernstein came about in response to the contemporary house style of *Artforum* which was 'widely regarded as the most theoretically advanced publication',[35] yet which also often seemed to sacrifice explanation for erudition in its defence of contemporary art.

Cheryl Bernstein made her first appearance in print as the author of the article, 'The Fake as More', which was a review of the work of the unknown but up-and-coming artist Hank Herron. Bernstein described Hank Herron as:

> A white, Anglo-Saxon New Englander seeped in modernist art and criticism, and making a debut in one of those stylish and expensive galleries that could count on critical attention from the high-art press.[36]

Herron's work was said to consist of nothing more than the exact replication of every art work that Frank Stella had produced up to that point in time (illus. 16). Yet the work was completed by Frank Stella over a ten year period (from 1961–1971); whereas Herron had taken just a year to complete the process of duplication. They were perfect fakes. And all the more profound for being so as Bernstein helpfully explained:

> On second viewing, one begins to be more profoundly conscious of and recaptive to a radically new and philosophical element in the work of Mr Herron that is precluded in the work of Mr Stella, i.e., the denial of originality, both in its most blatant manifestation (the fake as such) and in its subtle, insouciant undertones of static objectivity (the telescoping of time.) [...]. Mr Herron's art is neither deepening nor broadening nor, if anything, joyous. On the contrary, it is surface, narrow, and, most especially, tragic, for one is forcefully reminded at

34 Both 'The Fake as More', and 'Performance as News' appear in the anthology of Carol Duncan's work, *The Aesthetics of Power* (Cambridge: Cambridge University Press, 1993).

35 Ibid.

36 Ibid., p. 212.

> every line and turn that it represents the ontological predicament of our time, indeed of every living being: inauthentic experience. They are, in a word, fakes.[37]

These appropriations of Stella were the first works on public display by Herron. Thus there were no other pieces from his existing 'oeuvre' with which to compare this work and it was impossible to distinguish the two artists. After all, Herron's style was Stella's style alone. In this act of naked appropriation Herron appeared to have taken Stella's famous statement 'what you see is what you see', and purified it by removing all authorial connotation. He thereby seemed to have carried out Stella's project more successfully than Stella ever could. Stella had attempted artistic non-representation[38] but had always fallen short due to his own factual presence and status. All his works in fact did symbolically refer to something beyond themselves: to Stella the artist and to Stella the 'oeuvre' (just as the proper name Rembrandt does for the Rembrandt system).

Hank Herron's works on the other hand, could not be by Stella the man because they had supposedly originated from the hand of Herron. They were thus free from the symbolic function of representing 'Stella' and referred instead merely to themselves – they were, when regarded visually, truly non-representational. And likewise neither could Herron be considered the author of the works as they were on visual terms alone utterly indistinguishable from the works by Stella. Herron had added nothing to the visual experience, had no creative input, and could not be considered the 'author' of the patterns in all but the most general sense. It seemed therefore that Herron had 'out-Stella-ed Stella', whose work from the 1960s was indicative of, according to Leider:

> A cold, smartaleck, humourless, methodicalness that showed up, in the paintings he finally exhibited at the Museum of Modern Art in 1960 like a slap in the

37 Cheryl Bernstein, 'The Fake as More', in Duncan, *The Aesthetics of Power*, pp. 216–218.

38 For a fuller discussion of Stella and non-representation with reference to the concept of Literalism see Philip Leider, 'Literalism and Abstraction: Frank Stella's Retrospective at the Modern', in Frascina and Harris (eds.), *Art in Modern Culture, an Anthology of Critical Texts* (London: Phaidon, 1992), pp. 319–325.

face. The Paintings were insufferably arrogant. They seemed to reiterate nothing but insultingly simple principles that only his paintings, however, seemed to understand.[39]

Further despite (or maybe even because of) the cool irony of Herron's gesture it didn't seem incredible that such an artistic action could have taken place. One only needs compare Herron's gesture with that of the actual practice of artists such as Mike Bidlo, Malcolm Morley or Sherrie Levine all of whom have used a strategy of appropriating images (such as Picasso, Vermeer and Walker Evans) in their work.[40]

In Herron's case this lack of the actual objects of his work only served to heighten the overall credibility of him as a figure, as Duncan explained:

> Indeed, so fully theorised was Herron's work, that its physical absence from the art world would become a positive asset in the elaboration of its meaning. Never at any time would anyone wish to see the actual work or even wonder where it could be glimpsed.[41]

Having been written as a parody Bernstein's (or Duncan's?) article was first circulated in photocopy form amongst various people until it came to the attention of Gregory Battcock. Battcock included the paper in *Idea Art* which was his collection of articles on conceptual art first published in 1973. Battcock thereby made a major contribution to the subsequent 'lock-in' around Bernstein/Herron which ensured their subsequent existence within the discursive systems of the art world.

39 Ibid., p. 322.

40 In the thorough discussion of Hank Herron given by Thomas Crow in 'The Return of Hank Herron: Simulated Abstraction and the Service Economy of Art', Crow discusses the resonance between the imagined practice of Herron and the actual practice of appropriation artists such as Levine and Sturtevant (an artist who actually did copy Stella's paintings) in relation to the internal economy of the art market in the 1980s. See Thomas Crow, *Modern Art in the Common Culture* (New Haven: Yale University Press, 1996).

41 Ibid., p. 212.

Soon people began to talk about Herron as if he were a real artist (which by now he was, in a certain sense, in the process of becoming). The systems of discourse were becoming 'attracted' (in the sense of an attractor within a complex system) to the figures of both Herron and Bernstein who became 'real' attractors within the complex system of art discourse and its subsystems of art criticism and art history.

In her later commentary on the affair Duncan mentioned that Bernstein had developed an autonomous identity independent of Duncan:

> The concept of authorship has become in any case difficult to sustain [...]. Bernstein said things for me that I could not easily say. She brought difficult truths to light and then buried them again. Certainly she owes much to me, but whatever her debt, her life unfolded quite independently of mine. She established her own relationships (admittedly, they were one way), *entered into the discourse of art criticism* and helped more than one individual find his or her own critical or artistic path... On the basis of only 2 published texts she became a minor celebrity, a marginal but significant critical force in the art world, and, recently, the object of intense critical study. At present, the literature about her stretches several time the total length of her collected oeuvre. More interesting than her writing per se is Cheryl Bernstein the narrative construct[42]

In summary of the above: in the figure of Herron two themes of this and the preceding chapter are brought together – the dematerialised art work and the dematerialised artist. Yet this did not mean that Herron as an historical figure did not exist within the discursive and historical representational systems of the art world. This serves to illustrate that in ontological terms the figures of both the artist and the critic are like that of the artwork itself – namely relative to the representative systems which they are both subordinate to and constitutive of.

42 Duncan, *The Aesthetics of Power*, p. 211.

In concluding I return to the two leading claims of this chapter. First: discourse *in general* displays certain systemic principles and can therefore be understood in terms of a systems-thinking approach as a complex representational system. Second: discourse on art *in particular* can equally be understood in these terms. This is to say that discursive systems are complex systems too; and they will therefore display behavioural patterns common to all complex representational systems.

Above, this correlation was demonstrated by two examples. First it was shown *that* it was precisely due to their distributed nature that the discursive systems of art were robust enough to accommodate the absence of the artist as a person who lived. Second it was shown *how* these distributed discursive systems could become, like other systems, 'locked-in' around certain concepts and behavioural patterns. In the above examples discursive systems of art had become 'locked-in' around certain narrative tropes, such as the method of 'art history of the proper name' or common expectations of typical biographical motifs within art history, and also around particular artists, such as Rembrandt and Herron (but not Nat Tate).

Underlying all of these arguments is the claim that by identifying discourse on art as a subsystem of the system of art one can recognise that it is isomorphic to other complex systems. More specifically, isomorphic behaviour can be identified across all of the complex systems which feature in this book. These include: the art work/art works; the art gallery/museum; and discourse on art. In fact, even the book itself constitutes a part of this whole as it operates according to the concepts of the discursive systems of art. It too is 'locked-in' around certain fundamental concepts (namely, art, art history, and systems theory).

In this way we are able to map the behaviour of one system onto other systems. Thus it becomes evident that the isomorphic artistic systems of distributed representation, display and preservation of the art world are all locked-in around the concept of art. The very ambiguity of the concept of art serves to highlight that the processes of distributed representation, differentiation, display and preservation take place across the various systems of art. Consequently the meanings of the concept of art in the sub-systems of the artwork, the gallery and art discourse are systematically related and cannot be understood in isolation from one another.[43]

Furthermore, Thomas Crow observed processes of positive feedback and 'lock-in' (as identified by W. Brian Arthur in financial markets) are at work in the art market too: 'While governments and large corporations have plainly acquired a stake in contemporary art, the financial risks are miniscule by comparison.' Both the film industry and professional sports, for example, 'demand elaborate forms of concentration, monopoly and state intervention in order to secure the enormous financial commitments involved'. According to Crow, this is not true of the fine arts.

> With their low costs of entry and potential for exponential returns, the fine arts seem closer in this respect to computer software, our era's most potent form of intellectual property. Art seems similarly well adapted to a world in which markets are completely internationalised, politics are subordinated to them, and economic exchanges, unconstrained by time and space, are expressed in disembodied, quasi-fictional forms of information transfer.[44]

Curiously, the economy of art was described by Crow in almost the exact same terms in which William Burroughs described the economy of the heroin market. This could be taken to mean that art, like heroin, is one of the most exemplary expressions of the commodity system. They are both 'things which don't need any advertising'.[45]

43 In addition these processes of differentiation, whilst related to art in the systems outlined above, are not unique but are also observable across a range of complex systems including those studied in biology, physics, economics and cognitive science etc.

44 Crow, 'The Return of Hank Herron: Simulated Abstraction and the Service Economy of Art', in Crow, *Modern Art in the Common Culture*, p. 79.

45 'Junk is the ideal product [...] the ultimate merchandise. No sales talk necessary. The client will crawl through a sewer and beg to buy [...]. The junk merchant does not sell his product to the consumer, he sells the consumer to his product. He does not improve and simplify his merchandise. He degrades and

Perhaps then positive feedback in cultural systems (manifested in cultural and systemic memory, in irreversibility and ultimately cultural inertia) can be understood as a type of systemic 'addiction'. Thus Crow's claim that, 'the art world's continual circulation of information and services has become a primary source of profit in itself',[46] explains how the economy of the art market has 'locked-in' around particular cultural paradigms and has become a system which is self-organising and self-perpetuating (which is to say, in systems theoretical terms, auto-poietic). Extending this analogy further the other systems of the artworld, including art discourse, are not only 'locked-in' around art they are addicted to it too.

That the discursive systems of art are locked-in around the concept of art is something which has come under scrutiny in the ongoing debates concerning visual culture studies and the reconfiguration of art historical practice; a debate which is still an unresolved exchange of claim and counter-claim and which will not be revisited here. What I will take from these ongoing debates however is the non-trivial observation that art history requires both a concept of art and of history in order to function in a recognisable form (distinct from other academic disciplines). As de Duve claimed:

> Whatever the history of art – most often indeed the history of styles – does, it postulates (when it doesn't simply prejudge) the continuity of its substance, the invariance of its concept, the permanence of its foundations, and the unity of its limits. You are a historian of art and this is why, even though (all things considered) you don't know what art is, as far as you are concerned its history must be cumulative. Despite changes, even despite revolutions, the history of styles, accumulated in the mass of things that humans have called art over the course of time, appears to you as a cultural heritage. It belongs, you say, to humanity,

simplifies the client. He pays his staff in junk. Junk yields a basic formula of "evil" virus: "The Algebra of Need" The face of "evil" is always the face of total need. A dope fiend is a man in total need of dope. Beyond a certain frequency need knows absolutely no limit or control. In the words of total need: "Wouldn't you?" Yes you would. You would lie, cheat, inform on your friends, steal, do "anything" to satisfy total need. Because you would be in a state of total sickness, total possession, and not in a position to act in any other way.' William Burroughs, *The Naked Lunch* (London: Flamingo, 1986).

46 Crow, 'The Return of Hank Herron', p. 82.

and this is why your discipline is humanist. It is made up of objects but also of relations between those objects, ties of filiation, hinges of influence through which history obeys its own causality, broken influences and new departures through which art renews itself, naked as on the first day [...]. As a humanist historian, you redefine this corpus historically: it is a patrimony. Its manifest heterogeneity gives way to its cumulative continuity, which is grounded in the fact that an essence called art maintains itself unchanged through a succession of avatars.[47]

Finally there is also a political dimension to the critique of such discursive 'lock-in' or addiction, which couches the arguments in terms of an equation of cultural inertia with hegemony and power. Donald Preziosi has argued that the attractor 'art' (which I have identified here as a concept around which the systems of the artworld are 'locked-in') is part of a more general language of dominant cultural forms and norms. Art is, in short, part of the 'Esperanto' of western hegemony and power relations. He stated:

On a global scale art has come to be a universal method of (re) narrativising and (re)centring 'history' itself by establishing a standard or canon (or medium, or frame) in or against which all peoples of all times and all places might be seen together in the same epistemological space.[48]

And it is through identifying such hegemony in terms of the principles of systems theory that the potential for an engagement with such power structures of differentiation and representation can be realised with force and with efficacy.

47 Thierry De Duve, 'Art was a Proper Name', in *Kant After Duchamp* (Cambridge, MA: MIT Press, 1996), pp. 8 f.

48 Donald Preziosi, 'Collecting/Museums', in Robert Nelson and Richard Shiff (eds.), *Critical Terms for Art History* (Chicago: University of Chicago, 1996), p. 289.

Bibliography

Aarseth, Espen, *Cybertext: Perspectives on Ergodic Literature* (Baltimore: Johns Hopkins University Press, 1997).

Ackoff, Russell, 'Systems, Organizations and interdisciplinary research', *General Systems Yearbook*, vol. 5, 1960 (Society for General Systems Research).

Alberge, Dalya, 'A Very Private Collector: Charles Saatchi, Britain's Leading Modern Art Collector Talks (but only a little) to Dalya Alberge', *The Independent* (March 3rd, 1992).

Alloway, Lawrence, 'Network: the Artwork Described as a System', *Artforum* (1972), reprinted in Alloway, *Network: Art and the Complex Present* (Ann Arbor: UMI Research Publishing, 1984).

Alpers, Svetlana, *Rembrandt's Enterprise* (London: Thames and Hudson, 1988)

Angyal, A., *Foundations for a Science of Personality* (Boston: Harvard University Press, 1941).

—— *Logic of Systems* (London: Penguin, 1969).

Armstrong, Jeremy, and Byrne, Paul, 'Artless: Fury at Child-Fingerprint Portrait of Monster Myra', *Mirror* (July 26th, 1997).

Arthur, W. Brian, 'Self-Reinforcing Mechanisms in Economics', in Phillip Anderson, Kenneth Arrow, David Pines (eds.), *The Economy as an Evolving Complex System* (Redwood City, CA: Addison-Wesley, 1988).

—— 'Positive Feedbacks in the Economy', in *Scientific American* (Feb. 1990).

Atkins, P.W., 'The Rose, The Lion, and the Ultimate Oyster', *Modern Painters*, vol. 2, 4 (Winter 1989/90).

Bal, Mieke, *Reading Rembrandt* (Cambridge University Press, 1994).

Banham, Reyner, 'Enigma of the Rue du Renard', *Architectural Review* 963, vol. CLXI (1977).

Barthes, Roland, *Elements of Semiology*, trans. Annette Lavers and Colin Smith (London: Jonathan Cape, 1967).

Bass, Thomas, *The Predictors: How a Band of Maverick Physicists Set Out to Beat Wall Street* (London: Penguin, 2001).

Bateson, Gregory, *Steps to an Ecology of Mind* (New York: Ballantine, 1972).

Baudrillard, Jean, 'The Beaubourg-Effect: Implosion and Deterrence', trans. Rosalind Krauss and Annette Michelson, *October* 18 (1981).

—— *The System of Objects* (orig. French 1968), trans. James Benedict (New York: Verso, 1996).

Bertalanffy, Ludwig von, 'The Theory of Open Systems in Physics and Biology', *Science*, vol. 111 (1950)

—— *General Systems Theory* (New York: Braziller, 1968).

—— 'Foreword', E. Lazlo, *Introduction to Systems Philosophy* (New York: Gordon and Breach, 1972).

Bickers, Patricia, 'Editorial', *Art Monthly*, 222 (Dec. 1998–Jan. 1999).

Bijvoet, Marga, *Art as Inquiry: Toward New Collaborations Between Art & Science* (Oxford: Peter Lang, 1997).

Bishop, Claire, 'Antagonism and Relational Aesthetics', *October* (110, Fall 2004).

Boden, Margaret (ed.), *The Philosophy of Artificial Intelligence* (Oxford: Oxford University Press, 1990)

—— *The Philosophy of Artificial Life* (Oxford: Oxford University Press, 1996) and

Bonabeau, Eric, et al., *Swarm Intelligence: From Natural to Artificial Systems* (Oxford: Oxford University Press, 1999).

Boucher, W.I., and Quade, E.S. (eds.), *Systems Analysis and Policy Planning: Applications in Defense. A report prepared for United States Air Force, Project Rand* (New York: American Elsevier Publishing Company, June 1968).

Boulding, K.E., *The World as a Total System* (Beverly Hills: Sage, 1985).

Bourdieu, Pierre, Alain Darbel, Dominique Schnapper, *The Love of Art: European Art Museums and Their Public*, trans. Caroline Beattie and N. Merriman (Cambridge: Polity Press, 1991).

Bourriaud, Nicolas, *Relational Aesthetics* (Dijon: Les Presses du Réel, 2002, orig. 1997).

Boyd, William, *Nat Tate An American Artist: 1928–1960* (Sidney: 21 Publishing Ltd., 1998).

Brand, Stewart, *How Buildings Learn: What Happens After They're Built* (London: Penguin, 1997).

Broomhead and Jones, 'Time-Series Analysis', in Fleischmann et al., *Fractals in the Natural Sciences.*

Broos, Ben, 'Fame Shared is Fame Doubled', in *Rembrandt Impact of a Genius* (Amsterdam: K. and V. Waterman, 1983).

Brown, Richard, *Biotica: Art Emergence and Artificial Life* (London: RCA, 2001).

Buchanan, Mark, *Ubiquity* (London: Wiedenfeld and Nicolson, 2000).

Bureau of Labor Statistics, *U.S. Department of Labor,* Occupational Outlook Handbook, 2002–03 Edition, *Systems Analysts, Computer Scientists, and Database Administrators* at: http://www.bls.gov/data/ home.htm (25/11/2003).

Burn, Gordon, 'I Want It, I Want It All, and I Want It Now', *The Guardian G2* (Dec. 7th, 1998).

Burnham, Jack, *Beyond Modern Sculpture: The Effects of Science and Technology on the Sculpture of This Century* (New York: George Braziller, 1968).

—— 'Real Time Systems', *Artforum*, vol. 8, no. 1 (September 1969).

—— 'The Aesthetics of Intelligent Systems', in *On the Future of Art* (New York: Viking Press, 1970).

—— *The Structure of Art* (New York: George Braziller, 1971).

—— 'Steps in the Formulation of Real-Time Political Art', in Kaspar Koenig (ed.), *Hans Haacke/Framing and Being Framed: 7 Works, 1970–5* (New York: New York University Press, 1975).

—— 'Systems Esthetics', *Artforum* (September, 1968); reprinted in Donna de Salvo (ed.), *Open Systems: Rethinking Art c. 1970* (London: Tate, 2005).

Burroughs, William, *The Naked Lunch* (London: Flamingo, 1986).

Carter, Rita, *Mapping The Mind* (London: Weidenfeld & Nicolson, 1998).

Ceruzzi, Paul, *A History of Modern Computing: 1945–1995* (Cambridge, MA.: MIT Press, 1998).

Chadwick, George, *A Systems View of Planning* (Oxford: Pergamon, 1971).

Chia, Sandro, 'Saatchi's "Dispersals,"' Letter to *Art News*, vol. 84, no. 7 (Sept. 1985).

—— Statement in 'Making Art, Making Money: 13 Artists Comment', *Art in America*, vol. 78, no. 7 (July 1990).

Chomsky, Noam, *Syntactic Structures* (The Hague: Mouton, 1957).

Cilliers, Paul, *Complexity and Postmodernity* (London: Routledge, 1998).

Coddington, James, 'No Chaos Damn It', in Kirk Varnedoe and Pepe Karmel (eds.), *Jackson Pollock: New Approaches* (MOMA, 1999).

Coveney, Peter, and Highfield, Roger, *Frontiers of Complexity* (London: Faber and Faber, 1995).

Crow, Thomas, *Modern Art in the Common Culture* (New Haven: Yale University Press, 1996).

Crutchfield, Farmer, Packard, Shaw, 'Chaos', *Scientific American*, 255 (December, 1986).

Cummins, Robert, 'The Role of Representation in Connectionist Explanations of Cognitive Capacities', in: William Ramsey et al (eds.), *Philosophy and Connectionist Theory*.

Danto, Arthur C., *The Transfiguration of the Commonplace* (Boston: Harvard University Press, 1983).

—— 'Pop Art and Past Futures', in *After The End of Art* (Princeton: Princeton University Press, 1997).

—— 'The Artworld' (1964), in Steven Madoff (ed.), *Pop Art: A Critical Anthology* (Berkeley: University of California Press, 1997).

Debord, Guy, *The Society of the Spectacle*, trans. Ken Knabb (London: Rebel Press, 2004).

Dickie, George, 'Defining Art', *American Philosophical Quarterly* (July 1969) pp. 253–56.

—— *Art and the Aesthetic: An Institutional Analysis* (Ithaca: Cornell University Press, 1974).

—— *Art and Value: Themes in the Philosophy of Art* (Oxford: Blackwell 2001).

Dickson, Paul, *Think Tanks* (New York: Ballantine Books, 1971).

Duncan, Carol, *The Aesthetics of Power* (Cambridge: Cambridge University Press, 1993).

Duve, Thierry de, 'Art was a Proper Name', in *Kant After Duchamp* (Cambridge, MA: MIT Press, 1996).
Edwards, P., *The Closed World: Computers and the Politics of Discourse in the Cold War* (Cambridge, MA: MIT Press, 1996).
Fernie, Eric (ed.), *Art History and its Methods* (London: Phaidon, 1995).
Fischer, P. and Smith, William J. (eds), *Chaos, Fractals and Dynamics* (New York: Marcel Dekker Inc., 1985).
Fleischmann, M., Tildesley, D.J., and Ball, R.C. (eds.), *Fractals in the Natural Sciences* (Princeton: Princeton University Press, 1999).
Fodor, J. and Pylyshyn, Z., 'Connectionism and Cognitive Architecture', *Cognition*, 28 (1988).
Fodor, J., *The Language of Thought* (New York: Thomas Y. Crowell, 1975).
Foucault, Michel, 'What is an Author' (1969), in Donald Preziosi (ed.), *The Art of Art History* (Oxford: Oxford University Press, 1998).
—— *The History of Sexuality; Vol. 1, an Introduction*, trans. Robert Hurley (London: Penguin Books, 1998).
Frascina and Harris (eds.), *Art in Modern Culture, an Anthology of Critical Texts* (London: Phaidon, 1992).
French, Robert M., 'Semi-distributed representations and catastrophic forgetting in connectionist networks', in *Connection Science* 4 (1992).
Fried, Michael, *Courbet's Realism* (Chicago: University of Chicago Press, 1990).
Gelder, Tim van, 'Compositionality: A connectionist variation on a classical theme', *Cognitive Science* 14 (1990) pp. 355–384 and 'Defining "distributed representation,"' in *Connection Science*, 4 (1992).
Gentleman, Amelia, 'Fakes leave art world in chaos', *The Guardian* (Feb. 13, 1999).
Gere, Charlie, "New Media Art," *The Art Book*, vol. 12, 2 (2005).
Gere, Charlotte, and Vaizey, Marina, *Great Women Collectors* (London: Harry N. Abrams, 1999).
Gigliotti, Davidson, 'A Brief History of RainDance' (2003) on the website for *Radical Software* http://www.radicalsoftware.org/e/index.html (10/09/2007).
Glaister, Dan, 'Saviour or Destroyer? Maybe Both', *The Guardian* (Oct. 21, 1998).
Glancey, Jonathan, *The Guardian* (19 Jan., 2000).
Gleick, James, *Chaos* (New York: Vintage, 1987).
Gombrich, Ernst, *Art and Illusion* (London: Phaidon, 1960).
Goodwin, Brian, *How the Leopard Changed its Spots* (London: Weidenfield and Nicolson, 1994).
Greenberg, Clement, 'The Crisis of the Easel Picture' (1948), *Art and Culture* (Boston: Beacon Press, 1965).
—— 'Towards a Newer Laocoon', (1940), in Charles Harrison and Paul Wood (eds.), *Art in Theory 1900–1990* (Oxford: Blackwell, 1992).
Habermas, Jürgen, 'Modernity – An Incomplete Project', in Hal Foster (ed.), *The Anti-Aesthetic* (Port Townsend: The Bay Press, 1983).

—— 'Modernity versus Postmodernity', trans. Seyla Benhabib, *New German Critique*, 22 (1981).
—— *The Philosophical Discourse of Modernity: 12 Lectures* (Cambridge, MA: MIT Press, 1987).
Haken, H., *Synergetics: An Introduction: Nonequilibrium Phase Transitions and Self-Organization in Physics, Chemistry, and Biology* (New York: Springer, 1978).
Halsall, Francis, 'Niklas Luhmann', in Diarmuid Costello and Jonathan Vickery (eds.), *Art: Key Contemporary Thinkers* (Oxford: Berg Publishing, 2007).
—— 'Open Systems: Rethinking Art C. 1970', *The Art Book*, 13.3 (Aug. 2006).
—— 'Open Systems: Rethinking Art C. 1970', *CAA Reviews* (2005).
Hansen, Mark, *New Philosophy for New Media* (Cambridge, MA: MIT Press, 2004).
Harrison, Charles, and Wood, Paul (eds.), *Art in Theory 1900–2000* (Oxford: Blackwell, 2000).
Hatton, Rita, and Walker, John, *Supercollector* (London: Ellipsis, 2000).
Haugeland, John, 'The nature and plausibility of cognitivism', in *Behavioral and Brain Sciences*, 1 (1978).
Hayles, Katherine, *Chaos Bound: Orderly Disorder in Contemporary Literature and Science* (Ithaca: Cornell University Press, 1993)
—— *How We Became Posthuman: Virtual Bodies in Cybernetics, Literature, and Informatics* (Chicago: University of Chicago Press, 1999).
Hayles, Katherine (ed.), *Chaos and Order* (Chicago: University of Chicago Press, 1991).
Heinich, Nathalie, 'The Pompidou Centre and its public: the limits of a utopian site', Robert Lumley (ed.), *The Museum Time Machine* (New York: Routledge, 1988).
Heylighen, Francis, 'Self-Organization, Emergence and the Architecture of Complexity', *Proceedings of the 1st European Conference on System Science* (AFCET, 1989).
—— 'What is complexity?' (Dec 9, 1996,) published online at the *Principia Cybernetica Web* http://pespmc1.vub.ac.be/COMPLEXI.html (29/09/07).
Higgins, Dick, 'Intermedia' (1966), reprinted in Donna De Salvo (ed.), *Open Systems Rethinking Art c. 1970* (London: Tate Publishing, 2005).
—— *Horizons, the Poetics and Theory of the Intermedia* (Chicago: Illinois University Press, 1984).
Hinton, G. (et al.) 'Distributed Representations', in D.E. Rumelhart and James L. McClelland (eds.), *Parallel Distributed Processing* (Cambridge, MA. MIT Press, 1986) pp. 77–109.
Hobbs, Robert, *Robert Smithson; Sculpture* (Ithaca: Cornell University Press, 1981).
Holly, Michael Ann, *Panofsky and the Foundations of Art History* (Ithaca: Cornell University Press, 1984).
Hoogeveen, Jos and Hans Würzner (eds.), *Ideologie und Literaturwissenschaft* (Amsterdam: Rodopi, 1986)

Horgan, Terrance, and Tienson, John (eds.), *Connectionism and the Philosophy of Mind* (Dordrecht: Kluwer, 1991).
Husserl, Edmund, *The Crisis of European Sciences and Transcendental Phenomenology* (Evanston: Northwestern University Press, 1970).
Johnson, Paul, 'An Obscene Picture and the Question: Will Decency or Decadence Triumph in British Life?', *Daily Mail* (Sept. 20th, 1997).
Kanerva, Pentti, *Sparse Distributed Memory* (Cambridge, MA: MIT Press, 1988).
Karmel, Pepe, 'Pollock at work: The Films and Photographs of Hans Namuth', Kirk Varnedoe with Pepe Karmel, *Jackson Pollock: New Approaches* (New York: The Museum of Modern Art, 1998).
Kauffman, Stuart, *The Origins of Order* (Oxford: Oxford University Press, 1993).
Kelly, Kevin, *Out of Control* (New York: Perseus Books, 1994).
Kennedy, James, and Eberhart, Russell, *Swarm Intelligence* (San Francisco: Morgan Kauffman, 2001).
Kooning, Elaine de, 'Subject; What, How or Who?' *Artnews* (April 1955) cited in Irving Sandler, *The New York School* (New York: Harper and Row, 1978).
Krauss, Rosalind, 'In the Name of Picasso', *October* 16 (Spring, 1981).
—— *A Voyage on the North Sea: Art in the Age of the Post-Medium Condition* (New York: Thames & Hudson, 2000).
—— 'Reinventing the Medium (art and photography)', *Critical Inquiry*, vol. 25, no. 2 (Winter, 1999).
Krippendorff, Klaus, *A Dictionary of Cybernetics*, an eighty page report dated Feb. 2, 1986, published online at Principia Cybernetica Web, the website of the Principia Cybernetica Project (PCP) http://pespmc1.vub.ac.be/Default.html (10/12/ 2007).
Landesman, Peter, 'A 20th Century Master Scam', *New York Times Magazine* (July 18th 1999).
Lardner, James, *Fast Forward* (New York: New American Library, 1997).
Lewin, Roger, *Complexity* (New York: Macmillan, 1992).
—— *Complexity: Life on the Edge of Chaos* (London: Phoenix, 1993).
Lippard, Lucy, *Six Years: the Dematerialization of the Art Object* 1966–1972 (Berkeley: University of California Press, 1997 [1974]).
Lister, David, 'How a British Novelist Fooled the US Art World', *Independent* (April 7th, 1998).
—— 'London Partygoers Revel in Hoax that took in New York's Smartest', *The Independent* (April 9th, 1998).
Lorenz, Edward, *The Essence of Chaos* (London: UCL Press, 1995).
Luhmann, Niklas, 'Deconstruction as Second-Order Observing', in *New Literary History*, 24, 4 (1993).
—— *Social Systems*, trans. John Bednarz and Dirk Baecker (Stanford: Stanford University Press, 1995).
—— 'Why Does Society Describe Itself as Postmodern?' in *Cultural Critique*, no. 30 (Spring 1995) pp. 171–186.

—— *Art as a Social System*, trans. Eva M. Knodt (Stanford: Stanford University Press, 2000).
—— *Theories of Distinction*, ed. by William Rasch (Stanford: Stanford University Press, 2002).
Lyotard, *The Postmodern Condition* (Manchester: Manchester University Press, 1976).
Mandelbrot, Benoit, 'Fractal Geometry: what is it, and what does it do?', in Fleischman et al., *Fractals in the Natural Sciences.*
—— *The Fractal Geometry of Nature* (San Franscisco: WH Freeman and Company, 1977).
Manovich, Lev, 'Avant-garde as Software', *Artnodes*, http://www.uoc.edu/artnodes/eng/art/manovich1002/manovich1002.html (22/03/2006).
—— *The Language of New Media* (Cambridge, MA: MIT Press, 2001).
Marr, Andrew, 'Throwing the Baby Out With the Bath Water?' *The Guardian Saturday Review* (Sat. Oct. 24th, 1998).
Mitchell, W.J.T., *What Do Pictures Want?* (Chicago: University of Chicago Press, 2005).
Mondrian, Piet, 'Pure Art and Pure Plastic Art' (1936), in Herschel Chipp (ed.), *Theories of Modern Art* (Berkeley: University of California Press, 1968).
Morelli, Domenico, *Italian Painters* (1890) in Fernie, Eric (ed.), *Art History and its Methods.*
Mouland, Bill, 'Invitation to Outrage', *Daily Mail* (Sept. 17th, 1997).
Naifeh, Steven, and White Smith, Gregory, *Jackson Pollock: An American Saga* (New York: Clarkson N. Potter, Inc., 1989).
Owens, Craig, 'Representation, Appropriation and Power', *Beyond Recognition* (Berkeley: University of California Press, 1992).
Parsons, Talcott, *The Social System* (Glencoe: Free Press, 1951).
Peitgen, Heinz-Otto and Peter Richter, *The Beauty of Fractals* (New York: Springer, 1986).
Peitgen, Heinz-Otto and Saupe, Dietmar (eds.), *The Science of Fractal Images* (New York: Springer, 1988).
Penrose, Roger, *The Emperor's New Mind* (Oxford: Oxford University Press, 1989).
Petitot, Jean, et al., 'Beyond the Gap: An introduction to Naturalizing Phenomenology', in Petitot et al. (eds.), *Naturalizing Phenomenology.*
—— (eds.), *Naturalizing Phenomenology: Issues in Contemporary Phenomenology and Cognitive Science* (Stanford: Stanford University Press, 2000).
Pinker, S. and Prince, A., 'On Language and Connectionism', *Cognition*, 28 (1988).
Plato, *The Republic*, trans. H.D.P. Lee (London: Penguin, 2003).
Podro, Michael, *The Critical Historians of Art* (New Haven: Yale University Press, 1982).
Preziosi, Donald, 'Collecting/Museums', in Robert Nelson and Richard Shiff (eds.), *Critical Terms for Art History* (Chicago: University of Chicago, 1996).
Prigogine, Ilya, 'Thermodynamics of Evolution'. *Physics Today* (November, 1972).

—— and Stengers, Isabella, *Order Out of Chaos: Man's New Dialogue with Nature* (New York: Bantam, 1984).
Putnam, Hillary, *Representation and Reality* (Cambridge, MA: MIT Press, 1988).
Quade, E.S. (ed.), *Analysis for Military Decisions* (Chicago: Rand McNally & Co., 1964).
—— and W.I. Boucher, *Systems Analysis and Policy Planning* (New York: American Elsevier Publishing Co., 1975).
Rakuschan, F.E., 'Art and Repugnance: Form as Anti-Form, Art as Anti-Art, Market as Anti-Market', published on the website of the *European Institute for Progressive Cultural Policies*, http://www.eipcp.net/diskurs/d05/text/rakuschan_kunst_en.html (15/10/2007).
Ramsey, W. et al. (eds.), *Philosophy and Connectionist Theory* (Hove and London: Lawrence Erlbaum Associates, 1991).
Rasch, Willaim, *Niklas Luhmann's Modernity* (Stanford: Stanford University Press, 2000).
—— and Wolfe, Cary (eds.), *Observing Complexity: Systems Theory and Postmodernity* (Minneapolis: University of Minnesota Press, 2000).
Rayner, Jay, 'The Fake's Progress', *Observer Review* (July 4th, 1999).
Rembrandt Research Project, 'A Descriptive Survey of the Signatures', *A Corpus of Rembrandt Paintings I.*
—— *A Corpus of Rembrandt Paintings I* (Stichting Foundation/ Rembrandt Research Project, 1982).
Rampley, Matthew, 'Systems Aesthetics: Burnham and Others', in: *Vector [e-zine],* b # 12 (2005); http://virose.pt/vector/b_12/rampley.html
—— 'Art as a Social System: The Sociological Aesthetics of Niklas Luhmann', in: *Telos* (2008).
Reynolds, Ann, *Robert Smithson: Learning from New Jersey and Elsewhere* (Cambridge, MA: MIT Press, 2003).
Rosenberg, Harold, 'American Action Painters', in Charles Harrison and Paul Wood, *Art in Theory 1900–1990* (Oxford: Blackwell, 1991).
—— and Jakob, Seymour Slive, Engelbert Kuile, *Dutch Art and Architecture 1600–1800* (New Haven: Yale University Press, 1966).
Rumelhart, D.E. and J.L. McClelland, 'On learning the past tenses of English verbs', in J.L. McClelland, D.E. Rumelhart, and The PDP Research Group (eds.), *Parallel Distributed Processing*, vol. 2 (Cambridge, MA.: MIT Press, 1986).
Salvo, Donna de, 'Where We Begin – Opening the System, c. 1970', in D. De Salvo (ed.), *Open Systems Rethinking Art.*
Sayre, Kenneth, *Cybernetics and the Philosophy of Mind* (London: Routledge and Kegan Paul, 1976).
Sevänen, E., 'Art as an Autopoietic Sub-System of Modern Society', *Theory, Culture and Society*, vol. 18 (1) (London: Sage, 2001).
Shanken, E., 'The House that Jack Built: Jack Burnham's Concept of "Software" as a Metaphor for Art', in *Leonardo Electronic Almanac*, 6 (Nov. 1998).

—— 'Art in the Information Age: Technology and Conceptual Art', in Michael Corris (ed.), *Conceptual Art: Theory, Myth and Practice* (Cambridge: Cambridge University Press, 2004).

Shannon and Weaver, *The Mathematical Theory of Communication* (Chicago: University of Illinois Press, 1963 [1998]).

Shnayerson, Michael, in *Vanity Fair* (November, 2003).

Short, Larry, 'The Aesthetic Value of Fractal Images', *British Journal of Aesthetics*, 31:4 (1991).

Skinner, Quentin, 'Meaning and Understanding in the History of Ideas', in *History and Theory*, 8 (1969) pp. 3–53.

Smithson, Robert, 'Some Void Thoughts on Museums', 1967, in Jack D. Flam (ed.), *Robert Smithson: The Collected Writings* (Berkeley: University of California Press, 1996).

—— 'A Provisional Theory of Non-sites' (1968), in Jack D. Flam (ed.), *Robert Smithson: The Collected Writings* (Berkeley: University of California Press, 1996).

Smolensky, Paul, 'Connectionism, Constituency, and the Language of Thought', in B. Lower and G. Rey (eds.), *Meaning in Mind: Jerry Fodor and his Critics* (Oxford: Blackwell, 1991).

—— 'On the proper treatment of Connectionism', *Behavioural and Brain Sciences*, 11 (1988).

Spielmann, Y., 'Intermedia in Electronic Images', *Leonardo*, 34.1 (2001).

Stallabrass, Julian, *High Art Lite* (New York: Verso, 1999).

Stich, Stephen P., Rumelhart, David E., and Ramsey, William (eds.), *Philosophy and Connectionist Theory* (Hillsday, NJ: Lawrence Erlbaum Associates, 1991).

Taylor, Richard, Adam Micolich and David Jonas, 'Fractal Analysis of Pollock's Drip Paintings', *Nature*, vol. 399, 3 (1999).

—— 'Fractal Expressionism', *Physics World*, vol. 12, 10 (1999).

Thorpe, S., 'Localized versus distributed representations', in M.A. Arbib (ed.), *Handbook of Brain Theory and Neural Networks* (MA.: MIT Press, 1995).

To the North: Paintings by John Schueler, Brochure to the Exhibition at City Art Centre, Edinburgh (5th July–27th September, 2003).

Usselmann, Rainer, 'The Dilemma of Media Art: Cybernetic Serendipity at the ICA London', *Leonardo*, vol. 36, no. 5 (Oct. 2003).

Uvarov, E.B., Chapman, D.R., and Isaacs, Alan (eds.), *The Penguin Dictionary of Science* (London: Penguin Books, 1993).

Varadarajan, Tunku, 'British Joke Falls Flat in New York', *The Times* (April 8th, 1998).

Varela, Francisco, *Autopoiesis: A Theory of Living Organisation* (Amsterdam: North Holland, 1981).

Voss, Richard, 'Fractals in Nature: From Characterization to Simulation', in Peitgen/ Saupe, *The Science of Fractal Images*.

Wardrip-Fruin, N., and Monftort, N., *The New Media Reader* (Cambridge, MA: MIT Press, 2003).

Waugh, Alexander, 'The Great Painting of Myra "Designed to Shock"', *Evening Standard* (July 25th, 1997).

Ways, M., 'The Road to 1977', *Fortune* (Jan., 1967).

Weiner, Norbert, *Cybernetics or control and communication in the animal or machine* (Cambridge, MA: MIT Press, 1948).

Wetering, Ernst van de, and Broekhoff, Paul, 'New Directions in the Rembrandt Research Project, Part I: the 1642 Self-Portrait in the Royal Collection', *Burlington Magazine*, vol. 138, no.1 (1996).

Wilson, Peter, and Nairne, Sandy, 'The Installation File – Conserving by Documenting', *International Symposium on the Conservation of Contemporary Art, 7–12 July 1980, Abstracts* (National Museums of Canada, 1980).

Wolfe, Tom, *The Painted Word* (London: Black Swan, 1989).

Wypijewski, JoAnn, *Painting by Numbers – Komar and Melamid's Scientific Guide to Art* (New York: Farrar, Straus and Giroux, 1997).

Zijlmans, Kitty, 'Kunstgeschichte als Systemtheorie', *Gesichtspunkte. Kunstgeschichte heute* (Berlin: D. Reimer Verlag, 1995).

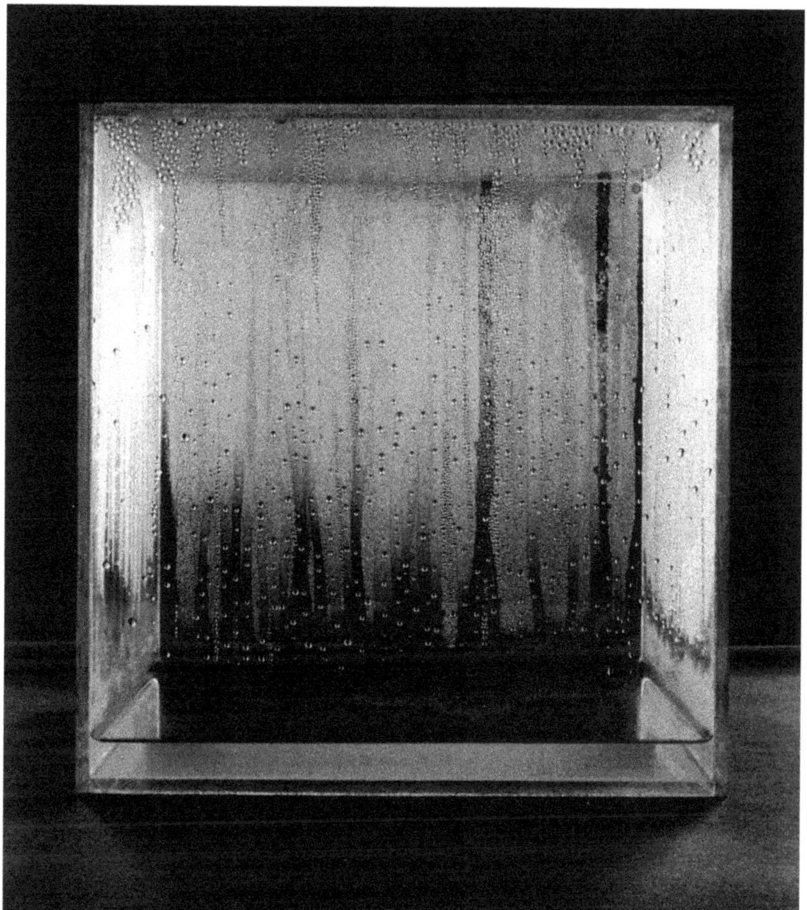

Illustration 1: Hans Haacke, *Condensation Cube* (1965), © DACS, London 2007

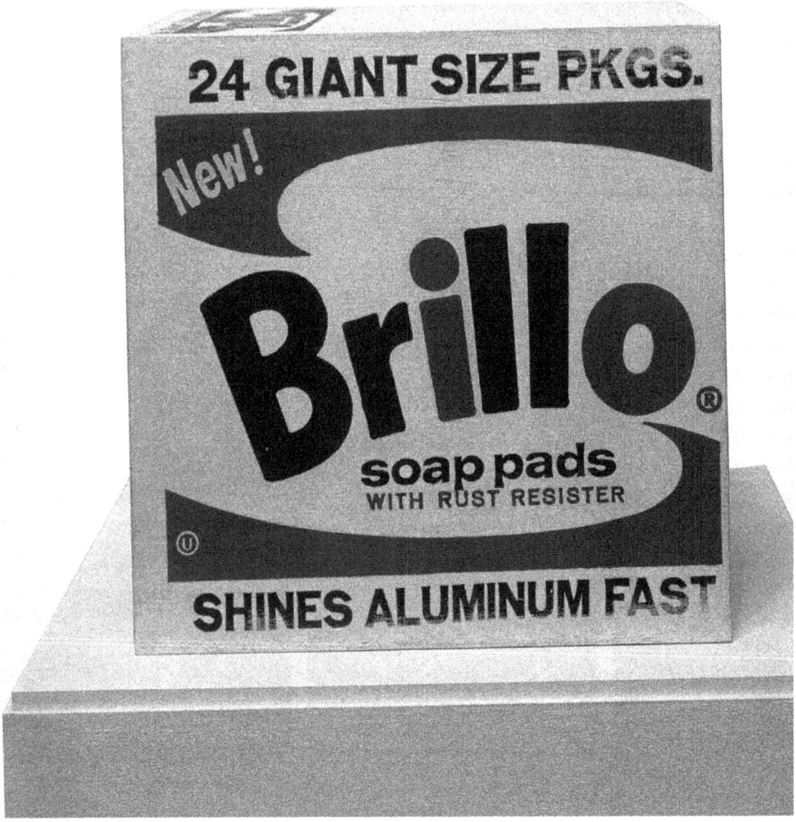

Illustration 2: Andy Warhol, *Brillo Box* (1964), © Licensed by the Andy Warhol Foundation for the Visual Arts, Inc/ARS, New York and DACS, London 2007

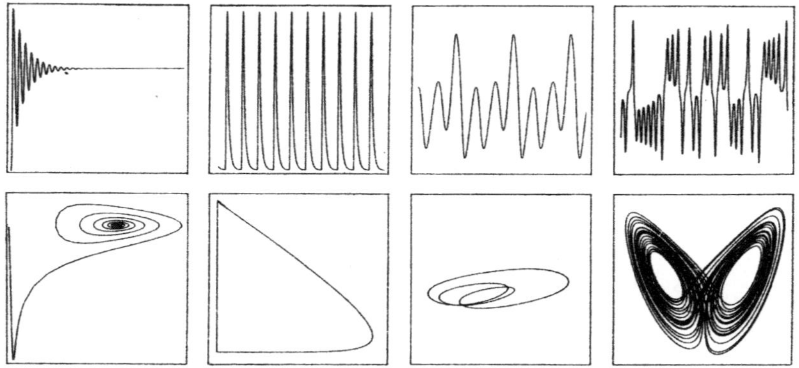

MAKING PORTRAITS IN PHASE SPACE. Traditional time series (*above*) and trajectories in phase space (*below*) are two ways of displaying the same data and gaining a picture of a system's long-term behavior. The first system (*left*) converges on a steady state—a point in phase space. The second repeats itself periodically, forming a cyclical orbit. The third repeats itself in a more complex waltz rhythm, a cycle with "period three." The fourth is chaotic.

Illustration 3: Image taken from James Gleick, *Chaos* (New York: Vintage, 1987)

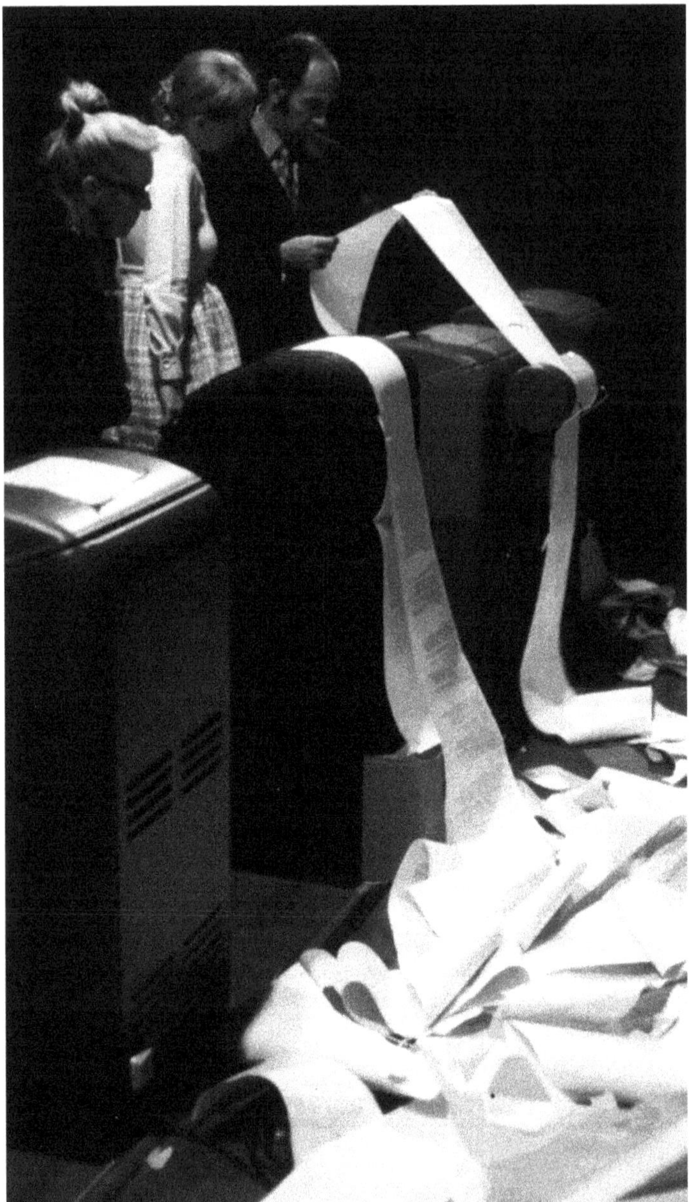

Illustration 4: Hans Haacke, *News* (1970), © DACS, London 2007

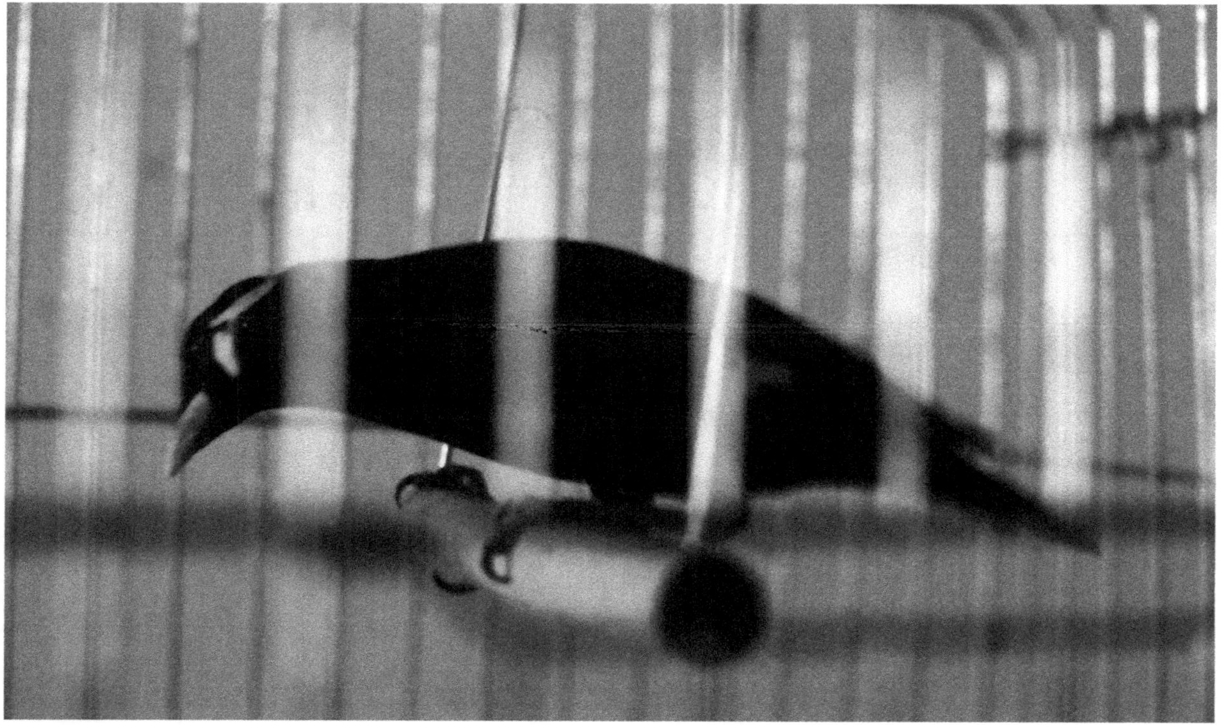

Illustration 5: Hans Haacke, *Norbert 'All Systems Go'* (1970), © DACS, London 2007

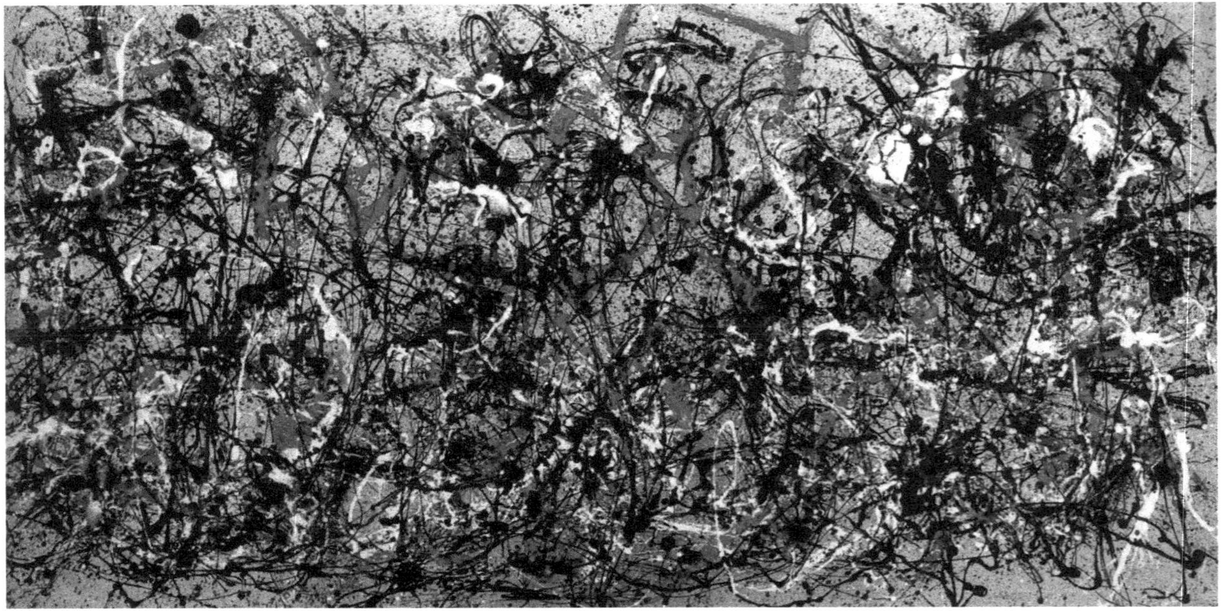

Illustration 6: Jackson Pollock, *Autumn Rhythm no. 30* (1950),
© ARS, NY and DACS, London 2007

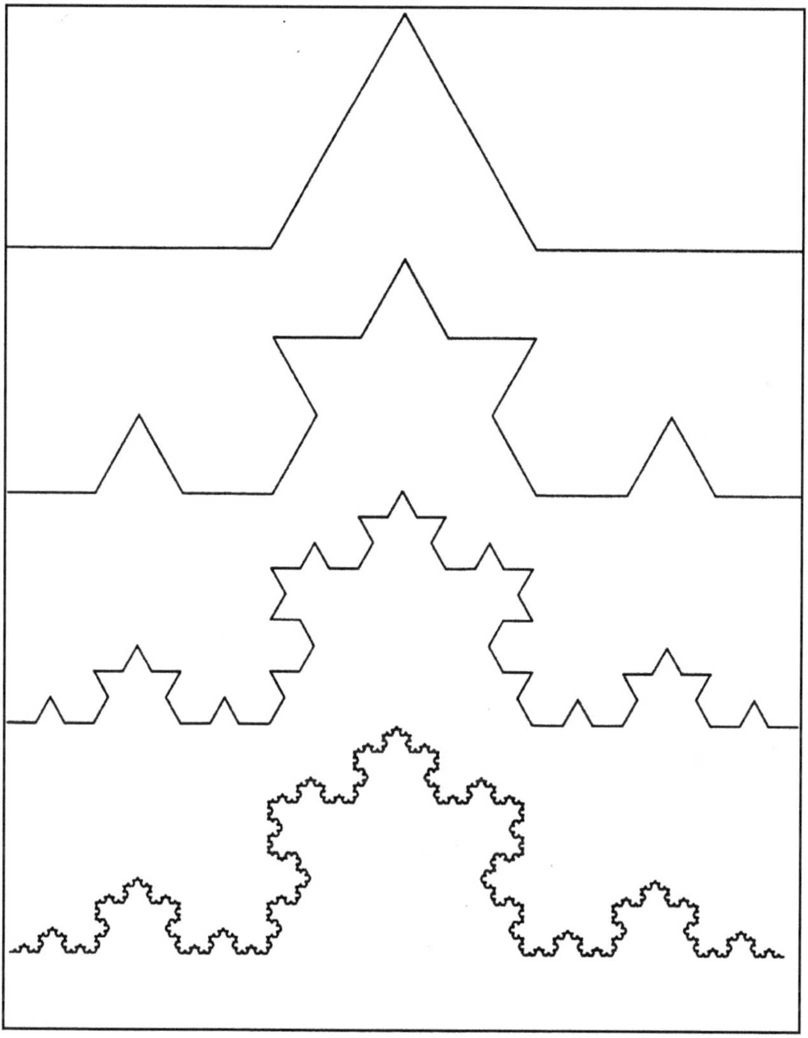

Illustration 7: A Von Koch Curve

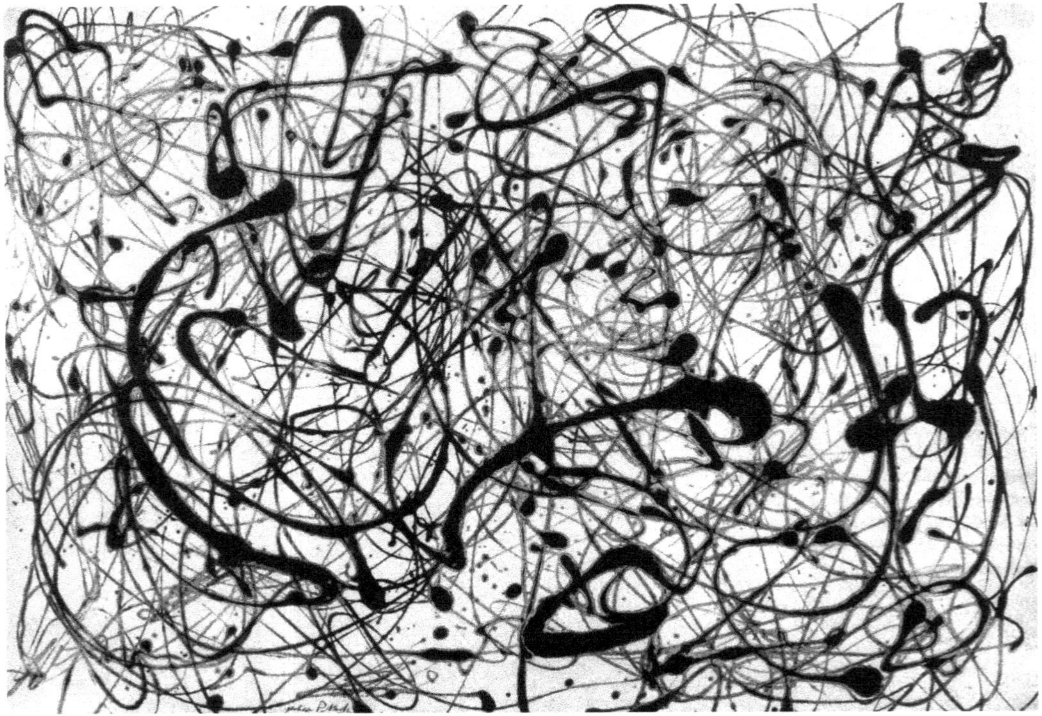

Illustration 8: Jackson Pollock, *No. 14* (1948), © ARS, NY and DACS, London 2007

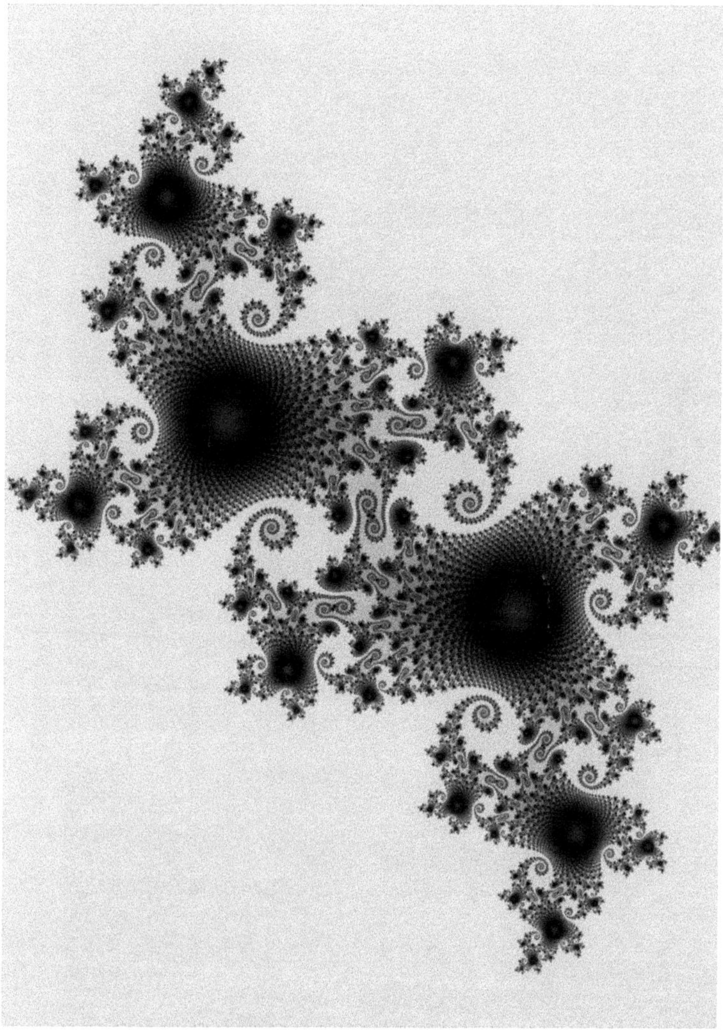

Illustration 9: A Julia Set Fractal, © Eequor

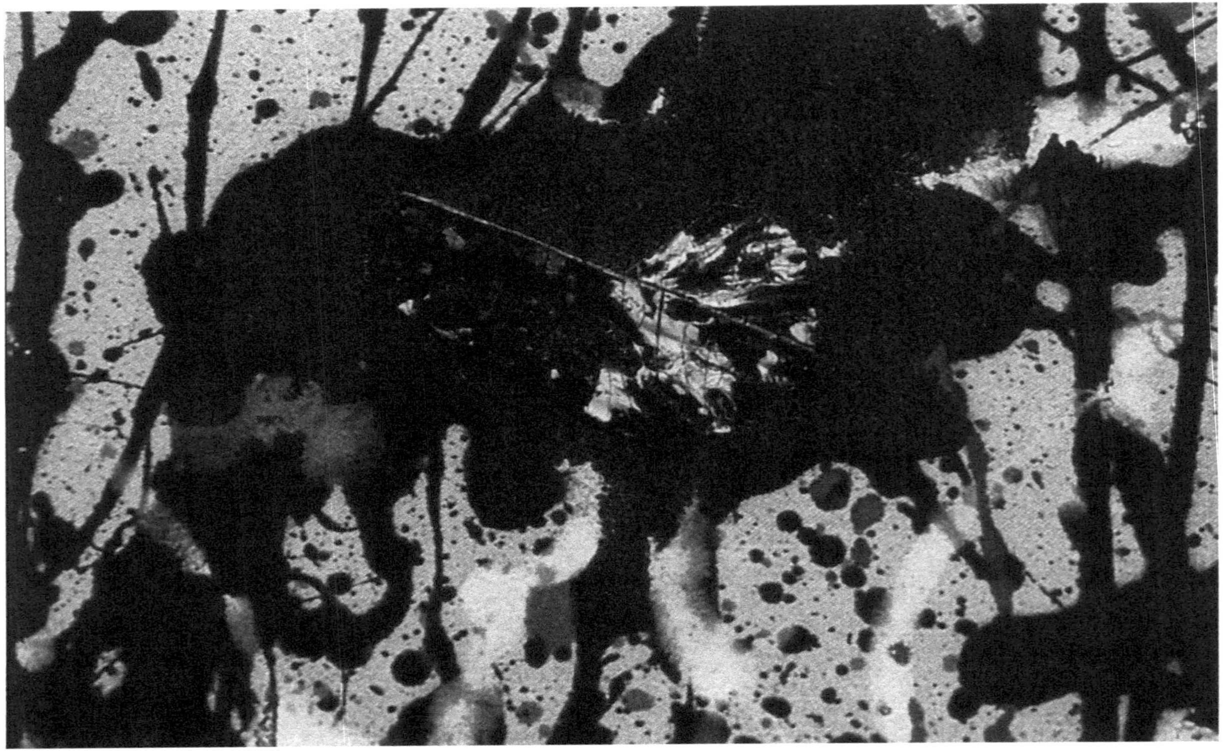

Ilustration 10: Jackson Pollock, *Autumn Rhythm no. 30* [detail] (1950), © ARS, NY and DACS, London 2007

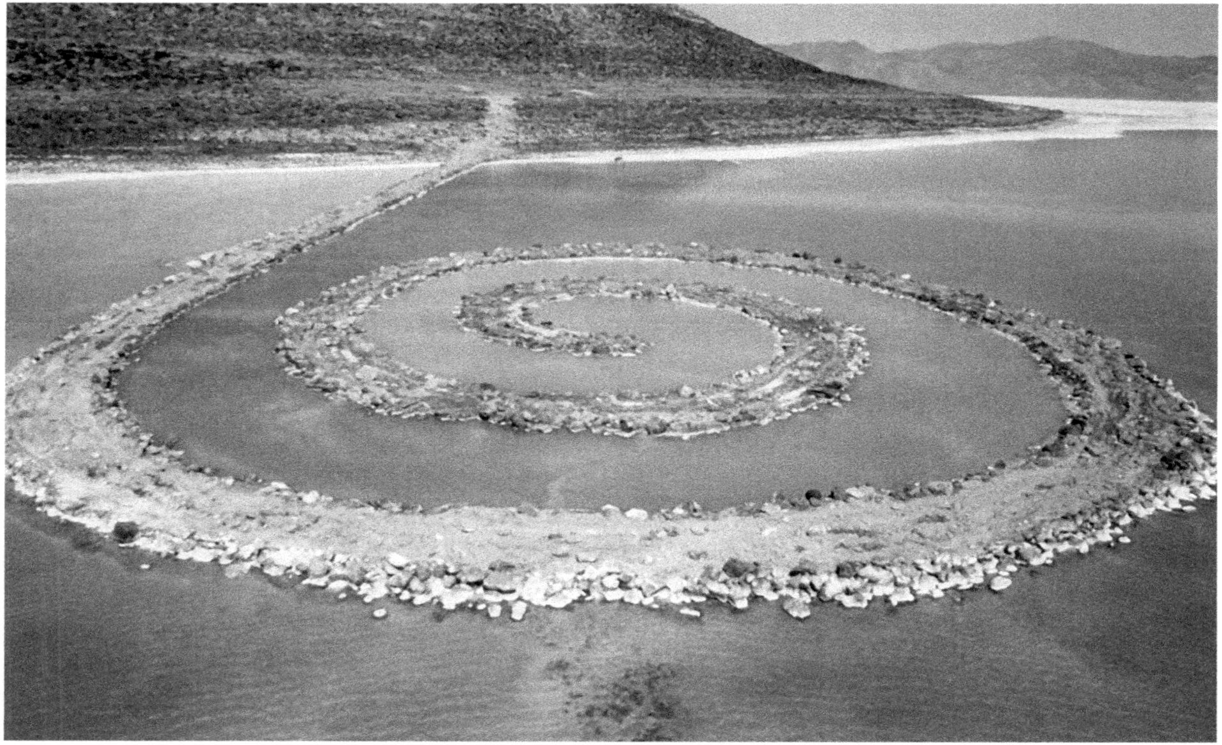

Illustration 11: Robert Smithson, *Spiral Jetty* (photograph by Gorgoni) (1970),
© Estate of Robert Smithson/DACS, London/VAGA, New York 2007

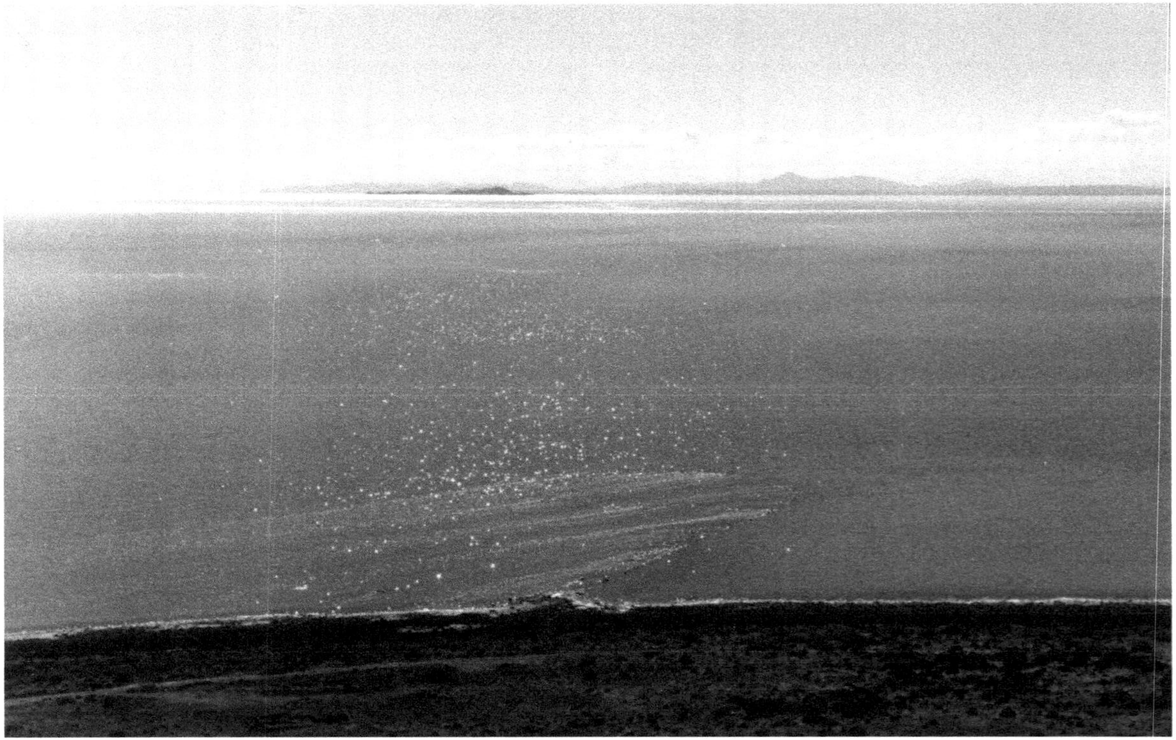

Illustration 12: Robert Smithson, *Spiral Jetty* (photograph by author, 2002)

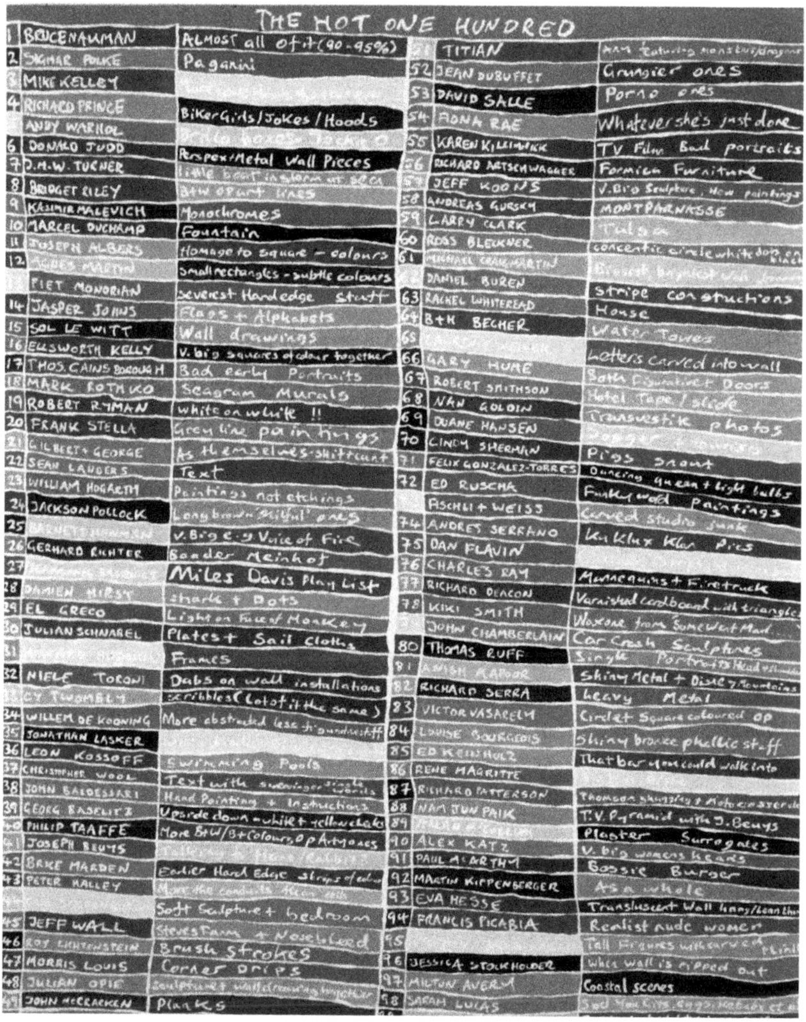

Illustration 13: Peter Davies, *Hot One Hundred* (1996),
© Peter Davies and Saatchi Gallery

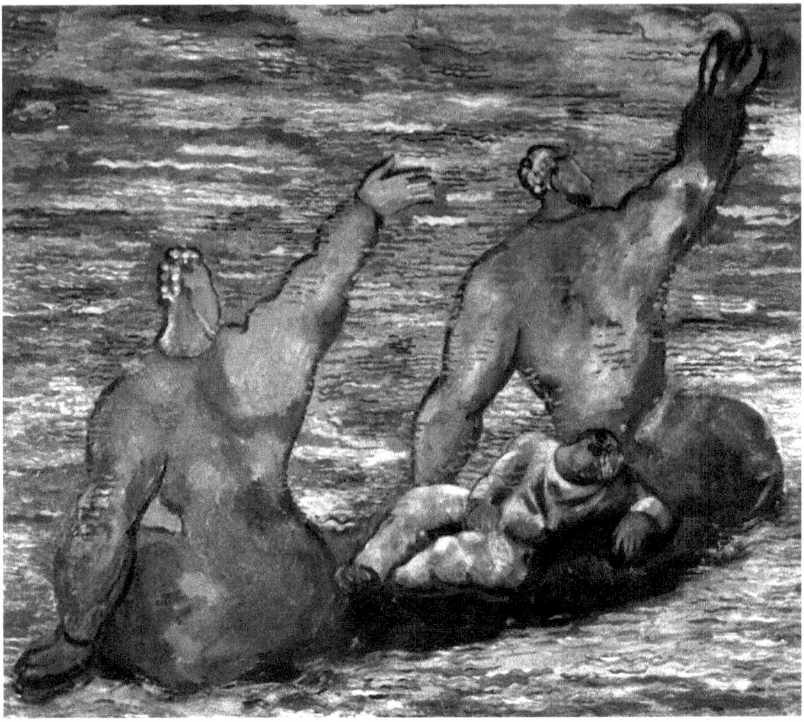

Illustration 14: Sandro Chia, *Three Boys on a Raft*, (1983),
© DACS, London/VAGA, New York, 2007

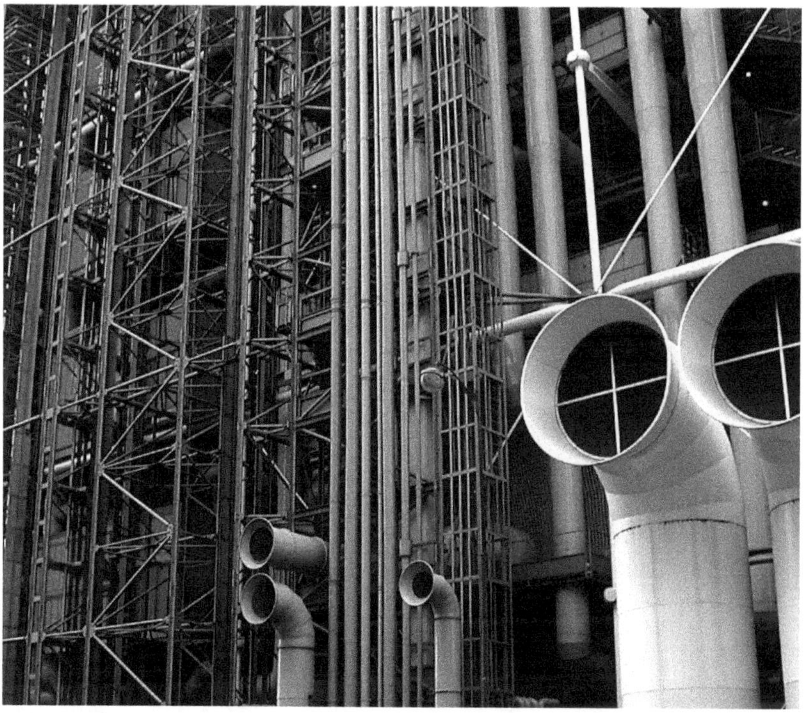

Illustration 15: An image of the Pompidou Centre

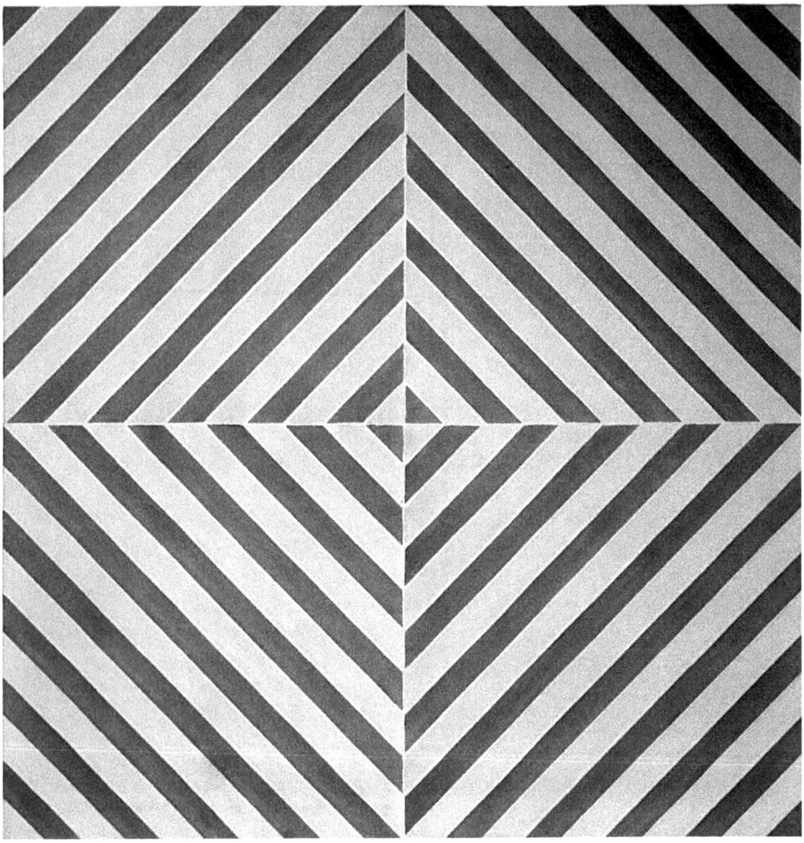

Illustration 16: Frank Stella, *Fez (2)* (1964),
© ARS, NY and DACS, London 2007

Index

www.ingramcontent.com/pod-product-compliance
Ingram Content Group UK Ltd.
Pitfield, Milton Keynes, MK11 3LW, UK
UKHW020719041026
15590UKWH00002B/768